Microwave and Radio Frequency Heating in Food and Beverages

Microwave and Radio Frequency Heating in Food and Beverages

Tatiana Koutchma

Research Scientist, Agriculture and Agri-Food Canada

Academic Press is an imprint of Elsevier
125 London Wall, London EC2Y 5AS, United Kingdom
525 B Street, Suite 1650, San Diego, CA 92101, United States
50 Hampshire Street, 5th Floor, Cambridge, MA 02139, United States
The Boulevard, Langford Lane, Kidlington, Oxford OX5 1GB, United Kingdom

Copyright © 2023 Elsevier Inc. All rights reserved.

No part of this publication may be reproduced or transmitted in any form or by any means, electronic or mechanical, including photocopying, recording, or any information storage and retrieval system, without permission in writing from the publisher. Details on how to seek permission, further information about the Publisher's permissions policies and our arrangements with organizations such as the Copyright Clearance Center and the Copyright Licensing Agency, can be found at our website: www.elsevier.com/permissions.

This book and the individual contributions contained in it are protected under copyright by the Publisher (other than as may be noted herein).

Notices

Knowledge and best practice in this field are constantly changing. As new research and experience broaden our understanding, changes in research methods, professional practices, or medical treatment may become necessary.

Practitioners and researchers must always rely on their own experience and knowledge in evaluating and using any information, methods, compounds, or experiments described herein. In using such information or methods they should be mindful of their own safety and the safety of others, including parties for whom they have a professional responsibility.

To the fullest extent of the law, neither the Publisher nor the authors, contributors, or editors, assume any liability for any injury and/or damage to persons or property as a matter of products liability, negligence or otherwise, or from any use or operation of any methods, products, instructions, or ideas contained in the material herein.

ISBN: 978-0-12-818715-9

For information on all Academic Press publications visit our website at
https://www.elsevier.com/books-and-journals

Publisher: Nikki P. Levy
Acquisitions Editor: Nina Bandeira
Editorial Project Manager: Emerald Li
Production Project Manager: Kiruthika Govindaraju
Cover Designer: Christian Bilbow

Typeset by TNQ Technologies

Contents

Introduction and Brief History of Microwave and Radio Frequency Heating

CHAPTER 1 Basic principles and mechanisms of electromagnetic heating technologies for food processing operations 3

1.1 Introduction 3
    1.1.1 Basic principles of electromagnetic heating technologies and their applications 3
    1.1.2 Microwave and radio frequency bands 8
    1.1.3 Mechanisms of microwave and radio frequency heat generation 11
    1.1.4 Microwave heating in food processing operations 14
    1.1.5 Radio frequency heating in food processing operations 23
1.2 Conclusions 25
References 26
Further reading 27

CHAPTER 2 Heating characteristics of microwave systems and dielectric properties of foods 29

2.1 Introduction 29
2.2 Microwave systems and their heating characteristics 29
    2.2.1 Design and operation principles of domestic and commercial microwave heating systems for food 29
    2.2.2 Heating characteristics of microwave systems 35
    2.2.3 Effect of product geometry 37
2.3 Dielectric properties of foods 38
    2.3.1 Effect of foods' dielectric properties on microwave heating 39
    2.3.2 Effect of food electrical conductivity on microwave heating 40
    2.3.3 Effect of other physical and thermal properties on microwave heating 40
2.4 Factors influencing dielectric properties of foods 41
    2.4.1 Temperature 41
    2.4.2 Moisture content 41
    2.4.3 Chemical composition 42

        2.4.4 Effects of nonelectrolytes in water 43
        2.4.5 Effect of pH and ionic strength.................................43
        2.4.6 Organic solids...43
        2.4.7 Proteins 43
  2.5 Propagation of microwave electromagnetic waves..................44
        2.5.1 Transmission properties of foods.............................44
        2.5.2 Wave impedance and power reflection 45
        2.5.3 Dielectric properties of mixtures..............................46
  2.6 Dielectric properties of foods in radio frequency range 49
  2.7 Conclusions..50
  2.8 Nomenclature 51
  References..51
  Further reading..53

CHAPTER 3   **Microwave heating effects on foodborne and spoilage microorganisms 55**
  3.1 Introduction..55
  3.2 Kinetics of microbial inactivation under microwave heating 57
  3.3 Effect of microwave energy on microorganisms.....................61
  3.4 Effects of food properties on microbial inactivation under microwave heating..64
  3.5 Effects on microwave heating parameters on microbial inactivation 66
        3.5.1 Microwave continuous flow systems.........................66
        3.5.2 Viscous products and products with particles 69
  3.6 Combined action of microwaves with other chemical or physical factors 73
  3.7 Conclusions..74
  References..75
  Further reading..78

CHAPTER 4   **Microwave heating and quality of food 81**
  4.1 Introduction..81
  4.2 Microwaves heating effects on overall quality of foods...........83
        4.2.1 Overall quality..84
        4.2.2 Moisture content 85
        4.2.3 Color..85
        4.2.4 Flavor 86
  4.3 Microwave heating effects on the destruction of vitamins and other nutrients in food......................................87
        4.3.1 Polyphenols...91

4.4 Microwave heating effects on lipids, proteins, and carbohydrates in foods 93
  4.4.1 Edible oils and fats 94
  4.4.2 Proteins 95
  4.4.3 Carbohydrates 96
  4.4.4 Polysaccharides 96
  4.4.5 Starch 97
  4.4.6 Microwave heating for developing foods with low glycemic index 98
  4.4.7 Minerals 98
 4.5 Microwaves heating and enzymes destruction in foods 99
 4.6 Effects of microwave heating on food chemistry 103
  4.6.1 Chemical reactions 103
  4.6.2 Acrylamide formation 105
  4.6.3 Thiobarbituric acid values 106
 4.7 Conclusions 107
 References 108

CHAPTER 5 Essential aspects of commercialization of applications of microwave and radio frequency heating for foods 113
 5.1 Introduction 113
 5.2 Packaging for microwave heating 114
  5.2.1 Passive packaging 114
  5.2.2 Active packaging 115
  5.2.3 Microwavable packages 116
 5.3 Microwave process validation 118
 5.4 Temperature and process lethality measurements during microwave heating 120
  5.4.1 Temperature probes 120
  5.4.2 Thermal imaging 121
  5.4.3 Process lethality indicators 121
 5.5 Microwavable foods and cooking instructions 122
 5.6 Regulatory status and commercialization 126
 5.7 Industrial microwave processes and systems 127
  5.7.1 Ready-to-eat meals and in-pouch sterilization 127
 5.8 Industrial radio frequency heating in processes and systems. 130
 5.9 Modeling of microwave heating systems 131
 5.10 Conclusions 134
 References 134

CHAPTER 6 Economics, energy, safety, and sustainability of microwave and radio frequency heating technologies ... 137

- 6.1 Introduction ... 137
- 6.2 Cost of capital equipment ... 140
  - 6.2.1 Operating costs ... 141
  - 6.2.2 Energy efficiency and cost ... 141
- 6.3 Savings from processing changes ... 144
- 6.4 Safety of microwave heating ... 146
- 6.5 Sustainability of microwave and radio frequency heating and equipment ... 147
  - 6.5.1 Microwave packaging sustainability ... 149
  - 6.5.2 Domestic microwave ovens ... 149
  - 6.5.3 Microwave-assisted technologies to achieve circular economy ... 150
  - 6.5.4 Microwave-assisted extraction ... 151
  - 6.5.5 Microwave-assisted pyrolysis of food waste ... 152
  - 6.5.6 Next development steps ... 154
- 6.6 Conclusions ... 154
- References ... 155

CHAPTER 7 Conclusions, knowledge gaps, and future prospects ... 157

Index ... 159

Introduction and Brief History of Microwave and Radio Frequency Heating

CHAPTER 1

Basic principles and mechanisms of electromagnetic heating technologies for food processing operations

1.1 Introduction

Five advanced heating techniques, infrared (IR), microwave (MW), dielectric or radio frequency (RF), ohmic (OH), and magnetic induction (MI) heating can heat foods faster and more efficiently because each of them utilizes various parts of electromagnetic (EM) energy spectrum. Volumetric or surface heating is generated in the product after absorption of the part of incident energy. The efficacy of these modes of heating is generally higher than that of conduction or convection heating modes. This chapter will briefly discuss the basic principles of available EM heating modes and their advantages and application. The focus will be given to dielectric heating modes using microwave and radio frequency (RF) energy and comparing differences in temperature distribution and conversion efficiency. Also, pros and cons of application of microwave and RF energy in food processing operations will be discussed including cooking, drying, extraction, tempering, thawing, blanching, preservation, and other new application including promising combined EM heating methods.

1.1.1 Basic principles of electromagnetic heating technologies and their applications

Microwave heating is a process within a family of advanced EM techniques. The main difference of advanced heating with conventional thermal processing systems is that the heat energy is transferred through conduction and convection from a hot medium to a cooler product that may result in large temperature gradients. Heat exchangers typically utilize pressurized steam from petroleum-fired boilers with less than 25%–30% of the energy conversion. Five advanced heating techniques, IR, MW, dielectric or RF, OH, and MI heating utilize EM energy and can heat foods faster and more efficiently. In the different electronic heating methods, it is important to recognize the interaction between the EM field at the frequency in question and the material being subjected to the energy. Except for MI heating, heat is generated within the product as a result of the transfer of EM energy directly into the product. This initiates volumetric heating due to frictional interaction between water molecules and charged ions. These methods offer a considerable speed advantage, particularly in solid foods and high efficiency of energy conversion ranging from 60% up to almost 100%.

OH or electric resistant heating relies on direct OH conduction losses in a medium and requires the electrodes to contact the medium directly. OH heating gives a direct heating because the product acts as an electrical resistor. The heat generated in the product is the loss in resistance.

OH heating devices consist of electrodes, a power source, and a means of confining the food sample (e.g., a tube or vessel) (Fig. 1.1). Appropriate instrumentation, safety features, and connections to other process unit operations (e.g., pumps, heat exchangers, and holding tubes) may also be important. OH heaters can be static (batch) or continuous. Important design considerations include electrode configuration (current flows across product flow path or parallel to product flow path), the distance between electrodes, electrolysis (metal dissolution of electrodes, particularly at low frequencies), heater geometry, frequency of alternating current (AC), power requirements, current density, applied voltage, and product velocity and velocity profile.

Since the heating effect depends on the eddy current induced in the material, this type of heating works well with conductors. In food processing, OH heating is used mainly for liquid products, as it is possible to establish the necessary electrical contact between electrodes and the media. The major benefits of OH technology claims include reducing heating time by 90%, uniform heating of liquids and liquids with particles with faster heating rates, reduced problems of surface fouling, no residual

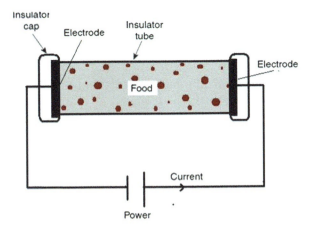

FIGURE 1.1

Schematic diagram of ohmic heating.

heat transfer after the current is shut off, low maintenance costs (no moving parts), and high-energy conversion efficiencies up to 97%–100%. OH technology is suitable for use in applications such as the processing of a low-acid particulate products in a can, meat cooking, stabilization of baby foods, and pasteurization of milk. Commercial OH heating systems are now available from a number of suppliers.

Magnetic induction heating produces heat by joule effect in a conductor by inducing eddy currents in a manner similar to that of a transformer when a current in the secondary windings is generated by induction from a current in the primary windings. Heating can be confined to that part of the work piece or material, which is directly opposite to the coil inducing the current (Fig. 1.2). In this system, the secondary coil is made of stainless steel pipe through which the products flow. Fluids are heated instantaneously as they pass through the pipe.

MI heats metal heat exchanger structures within the product flow stream. Inductively generated heat is transferred passively to the food and, unlike microwave or OH heating, is not produced in the food itself. Therefore, the United States Food and Drug Administration (US FDA) considers MI as a conventional thermal process. Since MI generates heat within metals, elaborate, even flow-driven rotary heat exchange surfaces become possible. Surfaces can be patterned to process foods of any viscosity or particulate composition. MI heat exchangers made of passive metal (such as 316 stainless) provide cheap, low-heat per-unit-surface-area heat

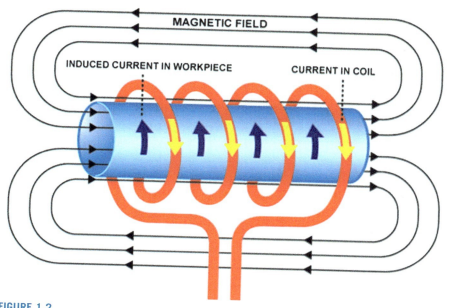

FIGURE 1.2

Schematic diagram of magnetic induction heating.

exchangers that potentially eliminate bake-on from high-protein/high-carbohydrate foods.

The induction coil and the heat exchanger do not have direct contact. The absence of direct utility contacts allows the MI heat exchanger to slip in and out of the induction coil with simple detachment of sanitary clamps. Therefore, swapping out heat exchangers for different food types or for cleaning becomes simple. MI can also heat makeup and cleaning water completely, eliminating the need for a boiler. MI heating can replace any industrial process currently serviced by steam. MI heating is all electric and because of this, MI allows zero carbon footprint food processing, if desired.

IR heating produced by EM radiation lies between the visible and microwave portions of the EM spectrum (Fig. 1.3).

IR light has a range of wavelengths (Fig. 1.4), just like visible light has wavelengths that range from red light to violet. "Near-infrared" (NIR) light is the closest in wavelength to visible light and "far-infrared" (FIR) is closer to the microwave region of the EM spectrum. FIR waves are thermal. NIR, mid-infrared, and FIR correspond to the spectral ranges of 0.75–1.4, 1.4–3, and 3–1000 μm, respectively.

As food is exposed to IR radiation, it is absorbed, reflected, or scattered. The amount of the IR radiation that is incident on any surface has a spectral dependence because energy coming out of an emitter is composed of different wavelengths and the fraction of the radiation in each band is dependent upon the temperature and emissivity of the emitter. The wavelength at which the maximum radiation occurs is determined by the temperature of the IR heating elements. When radiant EM

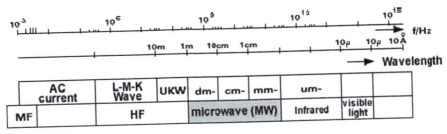

FIGURE 1.3

Electromagnetic radiation spectrum.

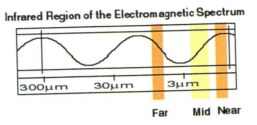

FIGURE 1.4

Infrared region of the EM spectrum.

energy impinges upon a food surface, it may induce changes in the electronic, vibrational, and rotational states of atoms and molecules. In general, the food substances absorb FIR energy most efficiently through the mechanism of changes in the molecular vibrational state. Water and organic compounds such as proteins and starches, which are the main components of food, absorb FIR energy at wavelengths greater than 2.5 µm.

1.1.1.1 Infrared heating food processing operations

Because IR's penetrating powers are limited, it can be considered as surface treatment in both liquid and solid foods. Recently, IR has been widely applied to various operations in the food industry, such as dehydration, frying, and pasteurization, as well as in domestic applications, such as grilling and baking. Electrical IR heaters are popular because of installation controllability, their ability to produce a prompt heating rate, and cleaner form of heat (Fig. 1.5). They also provide flexibility in producing the desired wavelength for a particular application.

The application of IR radiation to food processing has gained momentum. Recently, IR radiation has been widely applied to various thermal processing operations in the food industry such as dehydration, frying, and pasteurization. Even though IR heating is a promising novel method because it is fast and produces heating inside the material, its penetrating powers are limited. IR radiation can be considered as a surface treatment. Application of combined EM radiation and conventional convective heating is considered to be more efficient over radiation or convective heating alone, as it gives a synergistic effect.

IR heating can be used to inactivate bacteria, spores, yeast, and mold in both liquid and solid foods. Efficacy of microbial inactivation by IR heating depends on the IR power level, temperature of food sample, peak wavelength, and bandwidth of IR heating source, sample depth, types of microorganisms, moisture content, physiological phase of microorganism, and types of food materials.

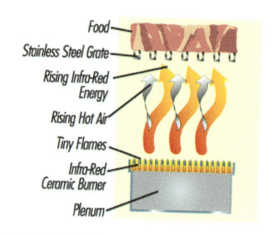

FIGURE 1.5

Schematic diagram of infrared heating.

Since the postprocess contamination of ready-to-eat (RTE) fully cooked meats primarily occurs on the surface, IR heating can be used as an effective postlethality intervention step that is necessary for meat processors to ensure the final microbial safety. According to the results of the USDA study (Huang and Sites, 2008), the IR surface pasteurization was effective in inactivating *Listeria monocytogenes* on RTE meats such as hot dogs. The pasteurization system contained four basic elements: an IR emitter, a hotdog roller, an IR sensor, and a temperature controller. The IR sensor was used to monitor the surface temperature of hotdogs. The IR emitter, modulated by a power controller, was used as a heating source. In all experiments, the temperature of the IR emitter was below 330°C, which can be easily achieved in the food industry. The hotdogs were surface-inoculated with a 4-strain *L. monocytogenes* cocktail to an average initial inoculum of 7.32 log (CFU/g). On the average 1.0, 2.1, 3.0, or 5.3 log reduction was observed after the surface temperature of hotdogs was increased to 70, 75, 80, or 85°C, respectively. Holding the sample temperature led to additional bacterial inactivation. With a 3 min holding at 80°C or 2 min at 85°C, a total of 6.4 or 6.7 logs of *L. monocytogenes* were inactivated. The combination of time and temperature was critical in IR surface pasteurization, as it is in all thermal processes. Without an additional holding period after come-up period, it was impossible to kill all bacteria if the original level of contamination was high.

In addition to the ability to combat Listeria and extend shelf life of cooked meat products, there is a growing interest in flame-broiling and simultaneously rapid cooking methods. Conveyorized IR broiling is a unique and innovative method based on medium wave carbon IR emitters that can heat meat surfaces in a targeted fashion. Sandwich meat, beef patties, hamburgers, and hams can be made to look even more appetizing without additional fat. Due to high temperatures and short cooking times, the IR broiler could produce more servings per hour compared to conventional gas heating.

In general, the operating efficiency of an electric IR heater ranges from 40% to 70%, while that of gas-fired IR heaters ranges from 30% to 50%. For the food sector, IR modules are manufactured in stainless steel and fitted with a wire mesh for mechanical protection. IR emitters can be switched on and off inside 1–2 s providing control from any unexpected or unwanted conveyor belt stoppage. Emitter failure detection is also incorporated within systems.

Being attractive primarily for surface heating applications, the combination of IR heating with microwave and other conductive and convective modes of heating holds great potential for achieving energy optimum and efficient practical applicability in the meat processing industry.

1.1.2 Microwave and radio frequency bands

Microwave frequency waves are generated through a magnetron applicator at frequencies between 300 MHz and 300 GHz, and, essentially, the interaction with the food material causes the food to heat itself (Fig. 1.6).

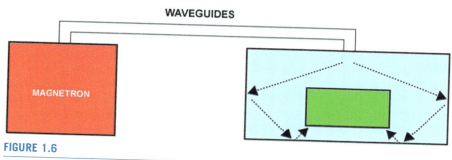

FIGURE 1.6

Schematic diagram of microwave heating.

Microwave energy is generated by special oscillator tubes, magnetrons, or klystrons; it can be transmitted to an applicator or antenna through a waveguide or coaxial transmission line. The output of such tubes tends to be in a range from 0.5 to 100 kW and requires a power supply. Microwaves are guided primarily by a radiation phenomenon; they are able to radiate into a space which could be the inside of the oven or cavity.

Microwave ovens incorporate a waveguide to deliver microwave energy to cook food in a cavity. In the microwave frequency range, the dielectric heating mechanism dominates up to moderated temperatures. The water content of the foods is an important factor for the microwave heating performance. For normal wet foods, the penetration depth from one side is approximately 1—2 cm at 2450 MHz.

The RF band of the EM spectrum (Fig. 1.1) covers a broad range of high frequencies, typically either in the kHz range (3 kHz < f < 1 MHz) or MHz range (1 MHz < f < 300 MHz). Both RF and microwave are considered to be a part of *nonionizing radiation* because they do not have sufficient energy (less than 10 eV) to ionize biologically important molecules or to break any chemical bonds.

Microwave and RF waves lie in the radar range and can interfere with communication systems; only selected frequencies are permitted for domestic, industrial, scientific, and medical applications. In the United States of America, these frequencies are 13.56, 27.12, and 40.68 MHz (RF), and 915 MHz, 2,450 MHz, 5.8 GHz, and 24.124 GHz (microwave). There are many distinct bands that are shown in Table 1.1 and used worldwide. Domestic microwave ovens commonly used for preparing RTE products operate at 2450 MHz, while most microwave industrial heating processes 915 MHz.

Simply placing a processed sample in a microwave heating system and expecting it to be heated efficiently is rarely fruitful due to the complexity of interaction of EM energy with food matrix. Nonuniformity of heating, difficulty to track cold spots, and unpredictable energy coupling are major issues when using microwave technology. However, rapid internal heating leaves the possibility of selective heating of materials through differential absorption and self-limiting reactions. This presents opportunities and benefits not available from conventional heating and provides

Table 1.1 Frequency allocation for industrial, medical, and scientific purposes worldwide[a].

Frequency, MHz	Frequency tolerance	Area permitted
433.92	0%–2%	Austria, The Netherlands, Portugal, Germany, Switzerland
896	10 MHz	UK
915	13 MHz	North and South America
2375	50 MHz	Albania, Bulgaria, Hungary, Romania, Czechoslovakia, Russia
2450	50 MHz	Worldwide except where 2375 MHZ is used
3390	0%–6%	The Netherlands
5800	75 MHz	Worldwide
6780	0%–6%	The Netherlands
24,150	125 MHz	Worldwide
40,680		UK

[a] Adapted from Metaxas, A.C., Meredith, R.H. Industrial microwave heating (IEE, 1983, reprinted 1988 and 1993).

an alternative microwave technology for a wide variety of products and processes. This includes defrosting, drying, blanching, pasteurization, and sterilization. In addition, microwave processing systems can be energy efficient up to 65%–70%. Heat can be instantly turned on and off—this constitutes important preconditions for the controllability of operation along with improvements in quality over conventionally processed products.

Food processors are creating a new generation of microwavable items due to consumers' increasing time pressure, nutritional awareness, and desire for foods that taste and smell like they were cooked in a conventional oven. In many cases, the products' success hinges on a combination of product reformulation, package design, and less processing. This new generation of microwave products includes meals, snacks, and everything in between, from fresh chicken and fish dishes, precooked entrees, and side dishes to grilled cheese sandwiches, biscuits, and pizza.

RF heating is also known as high frequency dielectric heating or "capacitive dielectric heating." During RF heating, the product to be heated forms a "dielectric" between two capacitor plates, which are then charged alternatively positive and negative by a high-frequency alternating electric field (Fig. 1.7).

Because RF uses longer wavelengths than microwave, EM waves in the RF spectrum can penetrate deeper into the product, so there is no surface overheating, hot or cold spots, a common problem with microwave heating. The RF heating also offers simple uniform field patterns as opposed to the complex nonuniform standing wave patterns in a microwave oven. RF heating involves the heating of poor electrical conductors. It is also characterized by freedom from electrical and mechanical contact

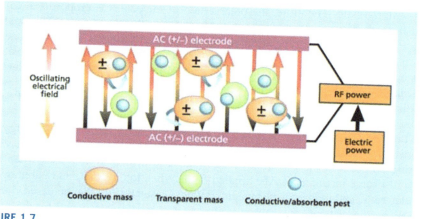

FIGURE 1.7

Schematic diagram of radio frequency heating.

with the food. RF heating involves the application of a high-voltage alternating electric field to a food sandwiched between two parallel electrodes. Typically, RF heaters generate heat by means of an RF generator that produces oscillating fields of EM energy and consists of a power supply and control circuitry, a hydraulic press and parallel plates, and a system for supporting processed material.

1.1.3 Mechanisms of microwave and radio frequency heat generation

Microwave and RF heating are accomplished as a combination of dipole rotation (i.e., when polar molecules try to align themselves in response to applied alternating electric field and interact with neighboring molecules, resulting in lattice and frictional losses as they rotate) and electric resistance heating resulting from the movement of the dissolved ions. Dielectric materials are composed of electrons, atoms, molecules, and ions. These constituents may be locked into regular structures of crystals or free to wander through the structure. Therefore, several types of electric displacements of these constituents such as electronic, atomic, molecular, and ionic are possible. The total displacement is the result of all types corresponding with the kinds of particle present in the material under consideration. The electronic displacement involves electrons moving with respect to the positive nucleus of each atom. The atomic displacement occurs when the atoms of each molecule move with respect to one another or with respect to the atoms in a crystal lattice. The molecular displacement involves molecules, which are initially asymmetrical in structure and have a definite electric moment in addition to their own electric field when placed in an external electric field. The uniform orientation of their axis is prevented by the forces of thermal agitation in the product. However, these forces are much smaller than that of the applied field which exerts a force on each molecule and aligns their axis with the field. Therefore, the polarization of molecules is proportional to the field strength. The ionic displacement is referred to as the "ionic drift." This applies only to heterogeneous but consists of two or more components or phases, one of which contains some free ions. An applied electric field will tend to push the positive ions in one direction and the negative ions in the opposite direction. The dielectric structure is such that a displacement brings a restoring force, which increases with the displacement until the restoring force just balances that due to the applied field. When the charges are driven back to their normal positions by the restoring force, and this force is no longer balanced by the applied field, the potential energy of the displaced charges is returned to the source: the free ions drift across the conducting particles in the direction of the field. The drift of ions ceases when the back EM force just balances the applied field. When the applied field is removed, the electronic and atomic displacements collapse instantly, but the increased charge caused by the drift of ions can only disappear at a rate determined by the ionic mobility.

The differences between temperature distributions, cold and hot spots inside the food for IR, OH electric, and high-frequency dielectric heating are shown below in Fig. 1.8.

The pros and cons of these five methods of EM heating are summarized in Table 1.2. This includes penetration depth, energy conversion, advantages, and limitations.

In summary, EM heating technologies offer local, all electrical heating solutions, with zero carbon footprint, and are highly efficient processing alternatives to steam with superior process control and optimal food quality. Despite these advantages, and the fact that EM techniques are available for many years, commercialization has been slow due to difficulties to implement expected benefits and lack of information on added benefits to convince food processors. Further studies are needed to investigate temperature distributions, develop appropriate process conditions with minimal quality losses, and design industrial system.

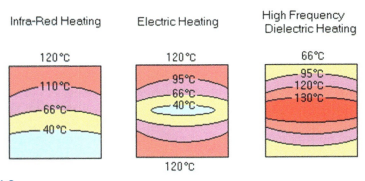

FIGURE 1.8

Temperature distribution for infrared, electric, and dielectric heating.

Table 1.2 Comparison of advantages and limitations of five electromagnetic heating technologies.

Infrared	Microwave	Radio frequency	Ohmic	Magnetic induction
Penetration depth Surface	Penetration depth depends on frequency and material	Penetration depth higher than microwave	Volumetric	Volumetric
Energy conversion, % 40–70	65–70	50–60	97–100	95
Best applications Baking Surface pasteurization	Continuous and in-package sterilization and pasteurization Viscous products	Insect control Defrosting Pasteurization and sterilization	Liquids with particulates	Liquids of any viscosity and with particulates
Packaging Prepackaging	Pre- and postpackaging	Pre- and postpackaging	Prepackaging	Prepackaging and metal packaging
Advantages High efficiency Fast Commercialized	Fast heating High efficiency Commercialized	Large objects	High efficiency and uniformity	Heats metal heat exchanger not the product
Limitations Low penetration depth	Nonuniformity	Low level of commercialization	Metal diffusion Not for all products	Low level of commercialization Limited research

1.1.4 Microwave heating in food processing operations

Microwave and RF heating offered a potential as alternatives to traditional processing operations such as preheating, baking, blanching, drying, defrosting, and pasteurization and sterilization for both solid foods such as prepacked RTE products, semisolid foods such as mashed potatoes and other prepared vegetables; and liquid foods such as milk and juice products, drinks, and other beverages. Microwave ovens became very popular home appliance that is used for reheating, preheating refrigerated RTEs, defrosting of frozen products, and cooking.

1.1.4.1 Microwave baking

There has been an increasing interest in the use of microwave heating and combination of microwave with other heating systems to reduce processing time and increase the quality of baked products. However, when the microwave technique was applied for bread cooking, there were a few main issues that the inside of bread was firm, while the outside was tough, and the low moisture content was observed. Several studies have been undertaken to improve the use of microwave energy to accelerate baking; however, a number of issues had to be addressed such as texture, low volume, lack of color, and crust formation, more dehydration, and rapid staling.

The use of susceptor lids manufactured by different companies is one commercial approach to solve these problems. Susceptors absorb microwave energy and convert it into heat which is transferred to the product by conduction or radiation creating localized areas of high temperature on the food surface. The temperature of the susceptor may reach 200—260 C.

In addition, it was reported that temperature of the dough at the dough susceptor interface in a microwave oven increased logarithmically with time and was a function of the weight and dimensions of the product. Browning of baked items was not observed without the use of susceptors made of microwave absorbing materials. Susceptors have been placed at the bottom of the sample to provide crust formation and surface browning of the food product.

Browning agents that develop a brown color on food surfaces during microwave heating are also available. Salt and coloring compounds, sometimes mixed with fats and sugars, have also been used to raise the surface temperature of the product to obtain some degree of browning. In addition, serious consideration should be given to dielectric properties and specific heat of food product. Specific heat is often a neglected parameter in microwave heating, but it should have an overriding effect especially for foods that have a low dielectric constant such as fats and oils.

A crisp crust and brown color have been obtained with microwave along with air jet impingement. Jet impingement baking is carried out by applying high speed convection with the help of commercial electrical oven. Microwaves were introduced from the top, and the air jets were introduced from both top and bottom at a velocity of 10 m/s. Microwave IR baking combined both microwave and IR heating. Microwave heating is not an ideal choice for bread-baking process, and more research and development is required to make it a commercially feasible option.

Lastly, traditional formulations of bread or bread-like doughs develop unacceptable textures when baked in the microwave oven. Microwave bread products with acceptable texture can be achieved by manipulating the gluten protein network, the size and the swelling of starch granules, and the moisture level.

1.1.4.2 Microwave blanching

Blanching is generally used for color retention and enzyme inactivation such as peroxidase, polyphenol oxidase, lipoxygenase, and pectins, which is carried out by immersing food materials in hot water, steam, or boiling solutions containing acids or salts. Blanching is usually considered a preheat treatment before drying, freezing, or canning operations. Blanching is usually conducted for fruits and vegetables for the purpose of removing the peel, modifying texture. Microwave blanching of fruits, vegetables, and legumes also has been studied and was found effective in combination with steam and water to reduce processing times and have less detrimental effects on nutrients such as ascorbic acid and plants tissue. Differences in results of these studies may be attributed to differences in blanching times as well as differences in the microwave ovens used. Variation in power output from oven to oven contributes to differences in results of studies on the use of microwaves.

For example, microwave blanching of herbs such as marjoram and rosemary was carried out by soaking the herbs in a minimum quantity of water and exposed to microwaves (Singh et al., 1996). Microwave blanching was observed for the maximum retention of color, ascorbic acid, and chlorophyll contents than that of water and steam blanching. Microwave-blanched samples were found to have better retention of quality parameter than that of microwave-dried samples without blanching.

Polyphenol oxidase (PPO) enzyme inactivation can occur during microwave and hot water blanching of mushrooms. PPO enzyme is the major enzyme responsible for browning reactions in mushrooms. Microwave blanching requires shorter time of 20 s for the complete inactivation of enzyme compared to the hot water blanching that takes more than 6 min due to the enhanced thermal effect during microwave blanching. Nevertheless, the direct application of microwaves to whole mushrooms may damage the sample texture due to the internal vaporization of water.

1.1.4.3 Microwave drying

The advantages of microwave drying are in achieving faster drying rates and consequently improving the quality of some food products. The energy absorption level is controlled by the product's moisture that can be used for selective heating of interior parts of the sample containing water and without affecting the exterior parts. Microwave drying is considered very beneficial during falling rate period when the diffusion is rate-limiting, resulting in the shrinkage of the structure and reduced surface moisture content. However, in microwave drying, due to volumetric heating, the vapors are generated inside and an internal pressure gradient is developed which forces the water outside.

Combining microwave energy with other drying methods can improve the drying process efficiency as well as the quality of food products which is far better than using conventional methods only. Microwave-assisted air drying is one of the methods where hot air is combined with microwave heating in order to enhance the drying rate. Microwave can be combined with hot air in three different stages of the drying process. At the initial stage, microwave is applied at the beginning of the dehydration process, in which the interior gets heated rapidly. At rapid drying period, a stable temperature profile is established in such a way the vapor is forced outside due to improved drying rate. This creates porous structure called as "puffing." It can further accelerate the mass transfer of water vapor. At the reduced drying rate period or at the final stage of drying, the drying rate begins to fall where the moisture is present at the center and with the help of microwave heating, vapor is forced outside in order to remove bound water (Zhang et al., 2006). Microwave-assisted air drying has been found to be helpful at the final stages of drying fruits and vegetables. Besides increasing the drying rate, microwave-assisted air drying enhanced the rehydration capacity of dried products and also overcame shrinkage problems.

During microwave-assisted vacuum drying, high-energy water molecules diffuse to the surface and evaporate due to low pressure. Because of this, water vapor concentrates at the surface and the low pressure causes the boiling point of water to be reduced. Vacuum drying prevents oxidation due to the absence of air, and thereby maintains the color, texture, and flavor of the dried products. In the absence of convection, either conduction or radiation or microwaves can be combined with vacuum drying to improve its thermal efficiency. Microwave-assisted vacuum drying has been applied for heat-sensitive materials such as banana, carrot, potato, etc. The loss of nutritional qualities (vitamins, α- and β-carotenes, etc.) of food products by microwave vacuum drying is minimum due to limited exposure to heat and oxygen.

Microwave freeze drying is considered as a gentle dehydration technique applied for heat-sensitive foods, pharmaceutical and biological materials (Zhang et al., 2006). In freeze drying, the temperature is reduced and, by applying vacuum or low pressure, the frozen water is directly transferred to the vapor phase without going through the liquid phase. Therefore, the pores of materials are preserved and those can be rehydrated quickly. The loss in terms of flavor can also be minimal using this method. Since freeze drying is time consuming, this method is applied only for high premium or heat-sensitive materials.

In microwave-assisted fluidized bed drying, the solid particles are forced to lift in the air stream, high rates of heat and mass transfer take place between the air and solid phases, and it is used for drying moist, granular materials (Ranjbaran and Zare, 2013). A continuous fluidized bed microwave paddy drying system was developed by Sangdao et al. (2010). This system was considered promising for paddy drying. The drying kinetics were studied of celery using a microwave-assisted fluidized bed dryer at 45°C, 55°C, or 65°C with 0 W/g, 1 W/g, and 2 W/g, respectively. It was shown that the model at 55°C and 1 W/g was the best for moisture ratio prediction, and the microwave energy reduced the drying time by more than 50% (Kaur et al., 2018). Compared to conventional fluidized bed drying, microwave drying techniques can greatly reduce the drying time of biological materials without quality degradation but has the disadvantage of nonhomogenous distribution in a microwave cavity, creating problems of nonuniform heating. Fluidized bed drying with microwave heating compensates for some of the drawbacks of each method. Temperature uniformity of the particles can be provided by good mixing due to fluidization, and the diffusion period of drying can be reduced by the utilization of microwave energy.

1.1.4.4 Microwave-assisted frying

Frying is a process of cooking foods in oil or another fat and to provide specific sensory qualities of products, including taste, texture, and color. Application of microwave energy in a frying system can reduce the processing time, change the internal pressure of the product rapidly, and maintain the oil quality. Microwave frying has two energy sources to heat foods, which can simultaneously cook the interior of foods by microwave and the surface of foods by hot oil. It achieves the necessary golden coloration and crispiness of the crust. Microwave frying should be made in a container made of a material that transmits microwaves. To achieve frying in a microwave oven, oil at room temperature is heated to the frying temperatures using microwaves. Then, the food is placed in heated oil and frying is carried out at the desired power level for a specified frying time. The effects of processing conditions on the quality of microwave-fried potatoes along with the effects of different oil types on the quality of fried products were studied. For all of the oil types, the increase in microwave power increased the moisture loss during frying. Osmotic dehydration is used as a pretreatment of conventionally fried foods to reduce the initial moisture content and as a consequence oil uptake. During osmotic dehydration, water diffuses out of the product, while the solute in the concentrated salt or sugar solutions diffuses into the product. Osmotic dehydration had a significant effect on oil and moisture content of French fries in microwave-assisted process. The application of osmotic dehydration reduced moisture and oil contents of fried potatoes. However, the hardness values of microwave-fried potatoes were higher than conventionally fried potatoes when osmotic dehydration was applied prior to frying.

Additionally, microwave vacuum frying (MVF) and vacuum frying (VF) techniques were compared for cooking potato chips. After the treatments, MVF significantly reduced the oil content in samples from 39.14 to 29.35 g oil/100 g (dry basis), increased the crispiness, and retained the natural color (Su et al., 2016). Modeling of the heat and mass transfer of water and oil during microwave frying of frozen potato chips has been investigated.

When testing the effects of microwave frying on acrylamide formation, it was reported that microwave frying provided lower acrylamide content and lighter color of French fries as compared to those fried conventionally for 5 min for all types of flours. This reduction in acrylamide level was the highest (34.5%) for rice flour containing batter.

Microwave frying is an alternative to conventional frying to improve the quality of fried potatoes and improve process efficiency. However, oil uptake in microwave-fried products may be lower or higher as compared with conventionally fried products depending on the type of the product, as in the case of VF. Another benefit is that potentially lower oil and acrylamide content in potato strips can be produced using microwave-assisted frying process.

1.1.4.5 Microwave-assisted extraction

Microwave-assisted extraction (MAE) is a heating physical method to increase the extracted bioactive compounds from foods or by-products. A solvent is used for compound extraction from a sample placed in the microwave zone, where the biomolecules and solvent vibrate with the alternating microwave fields. The examples of MAE were the extraction of polyphenol and caffeine from green tea leaves and by-products of wine production; pectin from potato pulp or mango peel and other valuable bioactive compounds. The results indicated that numerous benefits of the application of MAE included the following: process required shorter time due to internal heat and pressure generation, MAE provided the higher extractable yield of value-added compounds and was less labor-intensive. The extractable yield of pectin was 13.85% from mango peel using MAE at 700 W for 3 min, and the product had greater porosity compared to the conventional method (Wongkaew et al., 2020). MAE is considered green extraction technology.

1.1.4.6 Microwave-assisted puffing

Puffing of food is made by using high temperature, pressure, or extrusion. In puffing process, the expansion of seed is carried out and at that period the vapor pressure escapes through the micropores of the grain structure due to high pressure or thermal gradient. Explosion puffing is traditionally used for grains like millet, barley, black rice, rice, glutinous rice, and wheat and starch-based snacks. Microwave puffing is a new technology, which uses the characteristics of microwave heating to make the moisture in the material endothermic and vaporize, and then produce the starch gelatinization, protein denaturation, and moisture turn into steam in the material, thereby making the food material tissue puffing.

Microwave-assisted puffing has been recommended for low-fat products to replace deep-fried food products (Schwab et al., 1994). During microwave puffing, the material undergoes substantial structural changes and moisture loss. This process involves the complex physical phenomena of EM heating, heat and moisture transport, puffing, evaporation, and large levels of deformation (Rakesh and Datta, 2011).

Shrimp cassava cracker was developed by applying microwave puffing at 1200 W for 1 min. The resulting texture of the puffed microwave shrimp cracker was harder than for oil-fried cracker, whereas the volume expansion was less than that of oil frying. This might be explained by the formation of air bubbles inside the fried cracker caused by long frying and high temperature that create a more porous structure (Nguyen et al., 2013a, b). Addition of IR and hot air can potentially lead to better quality product, whereas using forced air convection is not desirable. Also, there is an optimum initial moisture content depending on the puffing conditions. The advantages of microwave puffing technology is in a fast-heating rate and a short heating time of the food. It is challenging to cause some unnecessary chemical reactions in the food with no increase in the fat of the food and retain the original product. The flavor makes the puffed food porous. Therefore, microwave puffing can have broad application prospects in the food industry.

1.1.4.7 Microwave tempering and defrosting

Tempering is the operation of thermal treatment of frozen foods to raise the temperature from below $-18°C$ to temperatures just below the melting point of ice (approximately $2°C$). For meats, such as steak or ground hamburger, the process of freezing then thawing permits further processing—from slicing and dicing to grinding and forming. Food processors portion servings by thawing frozen meats to a desired temperature at which they can then cut or split the product into fractioned amounts. A significant benefit of tempering is its ability to improve further processing downstream. Microwave tempering and defrosting are critical to the food processing industry. Since the 1970s, microwave energy has been applied to frozen food products, such as beef, poultry, fish, and sea products. After applying microwave heat, those who slice, form, dice, or grind food (particularly meat) will find a more uniformly conditioned product.

The process of "thawing" involves melting water crystals in the product. There are two conventional methods for tempering frozen product such as air and water-based processes. The process of tempering and thawing with industrial microwave systems saves time, space, energy, and proves to be beneficial as higher protein yields and flavor compounds remain in the product. This is a clean sustainable process that also saves water consumption and replaces labor-intensive cleanup and management. Microwave Techniques LLC (MT, USA) offers a range of customized industrial microwave systems for tempering and defrosting. Precise temperature control to $\pm 1°C$ and temperatures from $-18°C$ to $-3°C$ can be achieved in 3 min or less. Microwave tempering is no more expensive than air or water thawing when comparing actual costs. In fact, it is often less expensive when full cost accounting is properly employed.

1.1.4.8 Microwave pasteurization and sterilization

The uniqueness of microwave-assisted sterilization and pasteurization technology is that it can be applied to both solid and liquid foods as well as complete meals sealed in multicompartment trays. This includes sterilization, pasteurization, and shelf life extension of RTE meals or RTEs. Higher product quality can be achieved due to reduced processing times as compared to conventional thermal processes. The developers claimed that the application of microwaves could reduce the heating time of packaged foods from 1/4 to 1/10 of time required for conventional heating methods. Products for microwave sterilization have to be packaged in plastic trays or pouches. The ability of plastics to withstand oxygen permeation affects the organoleptic or sensory acceptance of the product during storage. With emerging innovative plastic technologies to the market, the new generations of plastics may increase the shelf life.

In 2011, the US FDA approved Microwave-Assisted Thermal Sterilization (MATS) Process using 915 MHz. As reported, the technology immerses packaged food in pressurized hot water while simultaneously heating it with microwaves at a frequency of 915 MHz. This combination eliminates food pathogens and spoilage microorganisms in just 5–8 min and produces safe foods with much higher quality than conventionally processed RTEs. Wornick Foods, a manufacturer of convenience foods and customized meal solutions, was awarded a grant to establish MATS Research and Development Center at its facility in Ohio where consumer products companies can test the MATS process.

915 Labs has the worldwide license to manufacture and market microwave-assisted sterilization and pasteurization technologies. According to 915 Labs, any food or beverage that will benefit from a lower processing temperature and a reduced processing time is ideal for MATS processing. Heat-sensitive foods such as eggs, dairy ingredients, seafood, and pastas have all been successfully processed with MATS. The reported research demonstrated that compared to conventional thermal retorting, the shortened exposure to heat during microwave processing retains more nutrients, such as Omega 3s, B vitamins, vitamin C, and folate.

The company currently manufactures a system that perform both sterilization (MATS) and pasteurization (MAPS) treatments. The pilot scale system is developed for product and process development. The smallest commercial production system processes 30 packages per minute. Larger capacity production systems with a throughput between 50 and 225 packages per minute are available by design. Actual capacities vary depending on the nature of the product, size of package, and the desired shelf life to be achieved. In order to realize the potential benefits of microwave processes and assist food companies to commercialize microwave technology and integrate it in the production process, 915 Labs offers product development, validation services, and packaging solutions.

Microwave continuous flow sterilization of homogeneous and particulate-containing high and low acid viscous products has been developed using frequency of 915 MHz. The US FDA acceptance has been also granted and cleared ways for commercial applications for pumpable low-acid foods.

The application of microwave pasteurization has been mainly applied to fluid foods such as pasteurization of juices and milk (2012). Microwave pasteurization of fresh juice and microwave sterilization of milk were reviewed (Salazar-Gonzalez et al., 2012). Microbial and enzyme inactivation in various beverages and semiliquid products was reported in products such as apple juice, apple cider, coconut water, grape juice, milk, and sweet potato puree. The knowledge of dielectric properties and product composition allowed to establish the appropriate processing conditions for applying microwave energy and achieve desired process lethality.

Pasteurization of in-shell eggs also can be achieved much faster and without negative impact on the shell integrity with the help of microwaves (Dev et al., 2008). It has been known that the albumen had higher dielectric properties than yolk. On contrary, microwave heating of in-shell egg did not show any significant difference in the heating rate of albumen and yolk. The enhanced interior heating might be due to the combination of egg geometry, dielectric properties, and size of the egg. It was reported that a 22% reduction of *Salmonella typhimurium* in the yolk of shell eggs (Shenga et al., 2010) was achieved for microwave heating during 15 s, whereas 36% reduction was achieved by moist heat treatment of 15 min.

1.1.4.9 Microwave-assisted freezing

Irreversible tissue damage caused by large ice crystals remains a challenge due to generated mechanical and biochemical stresses on the cellular structures of food materials during freezing. The application of electric and/or magnetic disturbances has been identified as a possible means to reduce the size of ice crystals. Xanthakis et al. (2014) added microwaves during freezing of pork meat. It was reported that the size of the formed ice crystals and damage of the microstructure of the meat was significantly greater in conventional freezing process as compared to freezing assisted by microwave energy. In particular, a 62% decrease in the average ice crystal size was observed. Generally, application of microwaves during freezing could reduce process time, lower energy consumption and optimize product quality.

1.1.4.10 Microwave-assisted infrared heating

As discussed earlier, application of IR radiation in food processing has been developed basically due to its benefits such as instant heating, rapid regulation response, low cost and compact equipment, and less changes in product quality. However, IR heating is regarded primarily as a technology for surface heating due to its low penetration in the product. Also, a prolonged exposure of food material to IR radiation could incur unwanted swelling and ultimately fracturing of the material. However, combining IR heating with microwave energy can be more profitable, since microwaves have greater penetration depth with volumetric heating and lower temperature gradient between the surface and interior of food material. Consequently, microwave-assisted IR process can overcome the limitations of IR heating, producing desirable foods to consumers.

1.1.4.11 Microwave-assisted infrared baking

The application of IR heating for the production of baked products and confectioneries has been limited, as a result of two factors: low in-depth heating and nonuniform moisture distribution within the heated food. However, a possible approach for overcoming this challenge is to combine microwave energy with IR heating, as microwave heating is associated with redeeming process time and ensuring an efficient distribution of temperature inside the product, while IR heating can significantly contribute to browning and crust formation. The promising results were obtained when process development was made along with the development of microwave-favorable product formulation and new product with improved health properties can result.

1.1.4.12 Microwave-assisted infrared drying

Infrared drying (IRD) has experienced growth of popularity as an alternative method for drying a wide range of agricultural products. However, as mentioned before, the main limitation of IR heating is its low penetrating depth; therefore, the synergistic action of microwave energy and IR radiation can resolve this drawback. It was reported that the combined drying technique had faster drying rates (time savings of approximately 98%) with lower final moisture contents (0.011−0.15 kg water/kg dry matter) for both banana and kiwi fruits when compared with conventional drying (Ozturk et al., 2016). In another study by Si et al. (2016), the drying kinetics and quality of raspberries under microwave-vacuum IR drying at different microwave powers and vacuum pressures were presented. By comparison, the drying process yielded 2.4 times better crispness value, 25.6% higher rehydration properties, 17.5% higher anthocyanin retention, and 21.2% higher DPPH radical-scavenging activity than single IRD at optimum conditions. Additionally, the process time for the combined microwave-vacuum IRD method was only 55% of IRD.

1.1.4.13 Microwave-assisted infrared roasting

Roasting is a pertinent process peculiar to some food products due to the enhancement of flavor, color, texture, appearance and accompanied with low investment cost and high production rates. However, literature reporting the utilization of microwave-infrared (MW-IR) combination ovens for roasting is very scarce. In a study by Uysal et al. (2009), an MW-IR oven was exploited for roasting of hazelnuts. Conventionally roasted hazelnut at 150°C for 20 min was used for comparison. Optimum roasting conditions were found to be only 2.5 min at 90% microwave power, upper and lower halogen lamps power at 60% and 20%, respectively, indicating that the roasting time of the hazelnuts was significantly reduced. Also, the MW-IR roasted hazelnuts had equivalent quality with conventionally roasted ones with respect to color, texture, moisture content, and fatty acid composition. Therefore, MW-IR combination oven had been recommended for roasting of hazelnut and other food products.

1.1.4.14 Microwave-assisted infrared tempering

In order to ensure that the desirable properties of food products have been preserved after tempering, MW-IR heating has been proposed as an energy-efficient and commercially feasible technique. For example, in the experimental investigation conducted by Seyhun et al. (2009), a halogen IR lamp-microwave combination oven was used to temper frozen potato puree. It was found that an increase in temperature and frequency caused significant changes in dielectric constant (ε') and loss factor (ε'') of the samples. In addition, tempering duration was substantially shortened as power levels increased. Therefore, microwave-assisted IR heating offers multitude benefits, which improves product quality and increases energy efficiency when compared to conventional heating techniques. Additionally, the prospect of commercial application of MW-IR heating is quite profitable.

1.1.4.15 Microwave-assisted ohmic heating

The combination of microwave and OH technique presents the possibility of heating food particles by microwaves without considering their electrical conductivity.

Besides, the liquid phase is heated via electric current and can improve heating uniformity. The feasibility of this techniques is in eliminating temperature lags in particulate food via the synergistic effect of microwave and OH combination mode using carrot cubes in 0.1% NaCl solution. When a combined two-step heating technique was employed, carrot cubes exhibited a temperature increase initially caused by microwave irradiation, and then this pattern was transposed during the second phase involving OH heating. Temperature difference between the carrot cubes and liquid eventually dropped to less than 2°C on the completion of the process in comparison to approximately of 8°C during separate heating processes Nguyen et al. (2013a, b).

Currently, most microwave-assisted food processing technologies are still at experimental phases, and there are very few or no reports on pilot or large-scale facilities. On the other hand, other emerging food processing technologies such as high-pressure processing, cold plasma, RF heating, ultraviolet light, pulsed electric fields, etc., could be combined with microwave energy and thus their feasibilities and effects should be tested.

1.1.5 Radio frequency heating in food processing operations

RF heating applications in the food industry has been recognized since the 1940s (Anonymous, 1993; McCormick, 1988). The main benefit of using EM energy in RF range with the longer wavelengths than microwaves is deeper penetration into the products, so there is no surface overheating, or hot or cold spots, which are common problems with microwave heating. The RF heating also offers simple uniform field patterns as opposed to the complex nonuniform standing wave patterns in a microwave oven. There are many applications of RF heating in the food industry because of their potential to improve the quality of food products and improve heating efficiency. The first attempts were to use RF energy to cook processed meat, to heat bread and cookies, and dehydrate vegetables. However, the primary commercial application in the late 1980s was the postbaking (final drying) of cookies and crackers. These RF systems have been recognized to be 70% efficient in removing moisture in comparison to 10% efficiency with conventional ovens (Memelstein, 1997). The MacrowaveTM 7000Series postbaking dryer developed by Radio Frequency Co., MA, generates heat inside of cookies (i.e., where the moisture content is greatest), resulting in a completely uniform moisture profile. This capability enables the baker to increase oven line productivity and develop the desired crumb structure, color, and target moisture content. Currently, RF postbaking drying is recognized as an advantageous processing technology for different types of cookies and crackers. The use of the MacrowaveTM postbaking dryer provides the modern baker an added tool to achieve the highest production yields obtainable without sacrificing quality. Other drying applications are increasing with an objective of disinfection of packaged flours as well as for drying granular foods with poor thermal characteristics such as coffee beans, cocoa beans, corn, grains, and nuts. Implementing a similar process, Radio Frequency Co worked closely with the Almond

Board of California to successfully develop a Macrowave Pasteurization System capable of providing a 5-log reduction of Salmonella in almonds, nuts, and other finished products and ingredients such as various types of flour, spices, xanthan gum, protein products, as well as pasteurization systems for animal feed products.

With major food recalls and the resultant implementation of the Food Safety Modernization Act (FSMA), the interest in inactivation and disinfestation applications using this mature technology has been grown and further developed.

RF heating is suitable for the sterilization and pasteurization of large packaged food. It was demonstrated that RF energy has adequate penetrations to inactivate heat-resistant bacteria spores in food prepackaged in 6-lb capacity polymeric trays. In RF sterilization system, the total time of processing in 6-lb trays could be reduced to 1/3 the time required in conventional canning processes to achieve approximately the same level of sterility for bacterial spores. Consequently, RF heating permits reaching better quality characteristics and improving consumer perception for a variety of products.

Research groups from Washington State University, University of Georgia (USA), and Seoul National University (Korea) have been investigating the potential of RF for pathogen control in bulk low-moisture foods and ingredients. Because of the low thermal diffusivity of low-moisture foods, conventional heating methods for bulk heating of low-moisture foods are slow and inefficient. Also, low water activity makes bacterial pathogens, e.g., Salmonella, extremely heat-tolerant. Using bench top-scale RF system for inactivation tests in wheat flour (water activity 0.25, 0.45, and 0.65), it was found that approximately 5–7-log Salmonella reduction can be achieved after ∼8.5–9 min of RF treatment. Furthermore, a 39-min RF treatment (6 kW, 27 MHz) was sufficient to inactivate Salmonella surrogates in 3-kg wheat-flour samples indicating that RF heating can be effective in inactivating pathogens in low-moisture conditions.

RF heating has been broadly researched for microorganism inactivation in various food commodities such as peaches/nectarines/stone fruits, ground beef, peanut butter cracker sandwiches, and packaged vegetables. Application of RF energy enhanced quality and improved process efficiency when used for liquid products pasteurization of cow and soy milk, orange and apple juice. Demeczky (1974) successfully showed that juices (peach, quince, and orange) in bottles moving on a conveyer belt through an RF applicator have better bacteriological and organoleptic qualities than juices treated by conventional thermal methods.

Thawing and tempering of frozen products were the next steps on the application of RF energy in the 1960s (Anonymous, 1993; Jason and Sanders, 1962).

Sairem of France and Stalam of Italy offer RF defrosting systems for thawing or tempering seafood, meat, and vegetables. Tempering raises the temperature of frozen blocks of fish products to just below the freezing point to facilitate cutting and further product handling. Their systems operate at a frequency of 27.12 MHz and output power from 10 to 105 kW. RF tempering reduced the time taken by traditional methods from tens of hours to 30 min, improved the homogeneity of product composition and thawing uniformity, showed a significant reduction of drip loss and micronutrient loss in RF-defrosted meat compared to air or water thawed product, reduced food waste and water use along with high-quality products for the consumer markets.

Insect infestation is another major area for applications of RF treatment. Chemical fumigants (like methyl bromide) are commonly used for postharvest pest control, but they have environmental concerns and regulatory issues. Alternative thermal treatments frequently result in product quality deterioration. RF technology allows for rapid and uniform heating of many substrates. Recent research has confirmed the viability of RF for the disinfestation of fruits and nuts.

1.2 Conclusions

Due to its speed and time-saving advantages over conventional heating, microwave heating has potential vast applications as pretreatment method and use in the food processing operations such as cooking, baking, drying, extraction, defrosting, pasteurization, and sterilization of foods. However, in food industry, the distribution of microwave-based processes is still in a relatively low numbers compared with their indisputable high potential. These successful microwave applications can range over a great spectrum of all thermal food processes. Reasons for the failure of implementations of industrial microwave processes range from high energy costs, which have to be counterbalanced by higher product qualities, over the conservatism of the food industry and relatively low research budgets, to the lack of microwave engineering knowledge and complexity of food systems. Also, challenges related to avoiding dielectric breakdown (arching) and thermal runaway heating from hot spots are being addressed by researchers and equipment manufacturers.

The future of microwave processing of foods appears to be the strongest for special applications, and it will probably be of limited usefulness as a general method of producing process heat. However, the greatest potential for microwave processing is in savings and improvements of productivity and a consequent decrease in labor and space costs. If improvements in quality, nutritional attributes, and taste over conventionally processed products are realized, the premium in price is obtainable for such advances. Thanks to these benefits, microwave energy represents today's heat processing method of choice.

Food processors are creating a new generation of microwavable items due to consumers' increasing time pressure, nutritional awareness, and desire for foods that taste and smell like they were cooked in a conventional oven. In many cases, the products' success hinges on a combination of product reformulation, package design, and less processing. This new generation of microwave products includes meals, snacks, and everything in between, from fresh chicken and fish dishes, precooked entrees, and side dishes to grilled cheese sandwiches, biscuits, and pizza.

Despite that RF technology has been available in food industry for many years, commercialization has been slow. Recently, because of the RF's unique benefits and high heating efficiency, this thermal processing technology has been extending its applications. Also, computer simulation has been an effective tool to study and better control electric field distribution and heating uniformity in complex food composition and advance design of RF systems for specific applications.

Lastly, RF and microwave heating are thermal processes caused by a nonionizing EM form of energy; it is not viewed as food additive. Therefore, microwaved or RF-treated products can carry the certified "organic" and "natural" label and doesn't require regulatory approvals. The microwave and RF pasteurization system can be implemented as a preventative control measure to provide a required kill step and produce better quality foods.

References

Anonymous, 1993. Radio frequency ovens increase productivity and energy efficiency. Prepared Foods 125. September.

Demeczky, M., 1974. Continuous pasteurisation of bot-tled fruit juices by high frequency energy. Proceedings of IV International Congress on Food Science and Technology IV, 11–20.

Dev, S.R.S., Raghavan, G.S.V., Gariepy, Y., 2008. Dielectric properties of egg components and microwave heating for in-shell pasteurization of eggs. Journal of Food Engineering 86, 207–214.

Huang, Sites, 2008. Elimination of *Listeria monocytogenes* on hotdogs by infrared surface treatment. Journal of Food Science 73 (1).

Jason, A.C., Sanders, H.R., 1962. Dielectric thawing of fish. Experiments with frozen herrings. Experiments with frozen white fish. Food Technology 16 (6), 101–112.

Kaur, A., Gariépy, Y., Orsat, V., Raghavan, V., 2018. Microwave assisted fluidized bed drying of celery. In: Proceedings of 21st International Drying Symposium. Valencia, Spain, pp. 1251–1260.

McCormick, R., 1988. Dielectric heat seeks low moisture applications. Prepared Foods September, 139–140.

Mermelstein, N.H., 1997. Interest in radiofrequency heating heats up. Food Technology 51 (10), 94–95.

Nguyen, T.T., Le, T.Q., Songsermpong, S., 2013a. Shrimp cassava cracker puffed by microwave technique: effect of moisture and oil content on some physical characteristics. Agriculture and Natural Resources 47, 434–446.

Nguyen, L.T., Choi, W., Lee, S.H., Jun, S., 2013b. Exploring the heating patterns of multiphase foods in a continuous flow, simultaneous microwave and Ohmic combination heater. Journal of Food Engineering 116 (1), 65–71.

Ozturk, S., Sakıyan, O., Ozlem Alifakı, Y., 2016. Dielectric properties and microwave and infrared-microwave combination drying characteristics of banana and kiwifruit. Journal of Food Process Engineering 40, e12502.

Rakesh, V., Datta, A.K., 2011. Microwave puffing: determination of optimal conditions using a coupled multiphase porous media–Large deformation model. Journal of Food Engineering 107, 152–163. https://doi.org/10.1016/j.jfoodeng.2011.06.031.

Ranjbaran, M., Zare, D., 2013. Simulation of energetic-and exergetic performance of microwave-assisted fluidized bed drying of soybeans. Energy 59, 484–493. https://doi.org/10.1016/j.energy.2013.06.057.

Salazar-Gonzalez, C., San Martin-Gonzalez, M.F., Lopez-Malo, A., Sosa-Morales, M.E., 2012. Recent studies related to microwave processing of fluid foods. Food and Bioprocess Technology 5, 31–46.

Sangdao, C., Songsermpong, S., Krairiksh, M., 2010. A continuous fluidized bed microwave paddy drying system using applicators with perpendicular slots on a concentric cylindrical cavity. Drying Technology 29, 35–46. https://doi.org/10.1080/07373937.2010.482721.

Schwab, E.C., Brown, G.E., Thomas, K.L., Harrington, T.R., 1994. High Intensity Microwave Puffing of Thick RTE Cereal Flakes. U.S. Patent No. 5,338,556. Filed February 23, 1993, and issue August 16, 1994.

Seyhun, N., Ramaswamy, H., Sumnu, G., Sahin, S., Ahmed, J., 2009. Comparison and modeling of microwave tempering and infrared assisted microwave tempering of frozen potato puree. Journal of Food Engineering 92 (3), 339–344.

Si, X., Chen, Q., Bi, J., Yi, J., Zhou, L., Wu, X., 2016. Infrared radiation and microwave vacuum combined drying kinetics and quality of raspberry. Journal of Food Process Engineering 39 (4), 377–390.

Singh, M., Raghavan, Abraham, K.O., 1996. Processing of marjoram (Majorana hortensis Moench.) and rosemary (Rosmarinus officinalis L.). Effect of blanching methods on quality. Food/Nahrung. 264–266.

Shenga, E., Singh, R.P., Yadav, A.S., 2010. Effect of pasteurization of shell egg on its quality characteristics under ambient storage. Journal of Food Science and Technology Mysore 47, 420–425.

Su, Y., Zhang, M., Zhang, W., Adhikari, B., Yang, Z., 2016. Application of novel microwave-assisted vacuum frying to reduce the oil uptake and improve the quality of potato chips. LWT–Food Science and Technology 73, 490–497. https://doi.org/10.1016/j.lwt.2016.06.047.

Uysal, N., Sumnu, G., Sahin, S., 2009. Optimization of microwave-infrared roasting of hazelnut. Journal of Food Engineering 90 (2), 255–261.

Wongkaew, M., Sommano, S.R., Tangpao, T., Rachtanapun, P., Jantanasakulwong, K., 2020. Mango peel pectin by microwave-assisted extraction and its use as fat replacement in dried Chinese sausage. Foods 9, 450. https://doi.org/10.3390/foods9040450.

Xanthakis, E., Le-Bail, A., Ramaswamy, H., 2014. Development of an innovative microwave assisted food freezing process. Innovative Food Science & Emerging Technologies 26, 176–181.

Zhang, M., Tang, J., Mujumdar, A.S., Wang, S., 2006. Trends in microwave-related drying of fruits and vegetables. Trends in Food Science & Technology 17, 524–534.

Further reading

De La Vega-Miranda, B., Santiesteban-Lopez, N.A., Lopez-Malo, A., Sosa-Morales, M.E., 2012. Inactivation of *Salmonella Typhimurium* in fresh vegetables using water-assisted microwave heating. Food Control 26, 19–22.

Metaxas, A.C., Meredith, R.H., 1983. Industrial Microwave Heating. IEE, 1983, reprinted 1988 and 1993.

CHAPTER 2

Heating characteristics of microwave systems and dielectric properties of foods

2.1 Introduction

Commercial applications of microwave and radio frequency (RF) energy for food heating and preservation include both in-pack processing of packaged foods (e.g., ready-to-eat [RTE] and ready-to-heat meals), heated continuously on a conveyor belt or as a batch, as well as in tubular in-flow preservation systems for fluid or semi-fluid pumpable foods based on high-temperature short-time processing. Domestic microwave ovens are broadly used around the world as kitchen appliance for fast reheating of refrigerated or frozen RTE products.

The system design, type of source of microwave energy, and process controls of microwave system define the heating performance of the microwave units in terms of its efficiency, heating uniformity, and safety. Additionally, the dielectric, thermal, physical properties, geometry and composition of the product impact greatly the absorption efficacy of microwave energy. This chapter will briefly introduce the operation and key elements of industrial and domestic microwave systems and discuss essential heating parameters that have to be known in order to evaluate the performance of microwave systems. Also, the dielectric properties of foods in microwave and RF ranges will be reviewed along with the effects of product chemical composition, moisture content, and other physical factors such as frequency, temperature, viscosity, and structure. Recent developments to advance smart microwave ovens will be presented.

2.2 Microwave systems and their heating characteristics

2.2.1 Design and operation principles of domestic and commercial microwave heating systems for food

Power supply, microwave energy sources, waveguides, and applicator chambers where the products are heated are key elements of any microwave heating systems.

Power supply draws electrical power from the line and convert it to the high voltage required by microwave energy sources.

2.2.1.1 Magnetron

Magnetron tubes serve as microwave energy source normally used for industrial and domestic applications that are capable of converting the power supplied into microwave energy. The magnetron usually requires several thousand volts of direct current (d.c). The cavity magnetron is a high-power vacuum tube with a central electron emitting cathode of high potential which is surrounded by a structured anode that generates microwaves using the interaction of a stream of electrons with a magnetic field while moving past a series of cavity resonators which are small, open cavities in a metal block. The anode structure forms cavities, which are coupled by their fringing fields and have the intended microwave resonant frequency. Due to the high electric d.c. field, the emitted electrons are accelerated radially but are deflected by an orthogonal magnetic d.c. field, yielding a spiral motion. If the electric and the magnetic field strength are chosen appropriately, the resonant cavities take energy from the electrons which can be coupled out by a circular loop antenna in one cavity into a waveguide or a coaxial line. The power output of a magnetron can be controlled by both the tube current and the magnetic field strength. The maximum power is generally limited by the temperature of the anodes. The realistic limits at 2.45 GHz are approximately 1.5 and 25 kW for air- or water-cooled anodes, respectively. Due to their larger size, 915 MHz magnetrons can achieve higher powers per unit. The efficiencies of modern 2.45 GHz magnetrons range at approximately 70% most limited by the magnetic flux of the economic ferrite magnets.

Magnetron-based microwave heating appliances are known for the instability of energy efficiency, low reproducibility, nonuniform temperature distribution in the load, and unpredictable microbial inactivation performance. Potential risk of foodborne outbreaks is linked to microwave heating of a variety of products.

2.2.1.2 Solid-state generators

Solid-state (SS) microwave and RF sources have been recently introduced as capable of performing electronic control over frequency and magnitude of the excited microwave field and heating processes. Using cutting-edge gallium nitride (GaN) semiconductors, microwave generator systems are developed and have a potential to provide optimal stability and controllable microwave power demanded by industrial heating applications. It is now acknowledged that the rapidly growing technology of generation of microwave energy by SS semiconductor chains may make a revolutionizing impact on the entire field of microwave power engineering. The reason for this is the potential capability of this technology to maintain electronic control over the key electromagnetic characteristics of microwave heating systems. The possibility of controlling frequency, magnitude, and phase of the signal is expected to essentially improve (in comparison with traditional magnetron generators) the controllability of microwave thermal processing in terms of both of its key characteristics: energy efficiency and distribution of heating patterns. Additionally, magnetrons have a limited lifespan and can lose their emitting properties after approximately 10,000 h. SS generators have a much longer service life that is almost unlimited. Unlike magnetrons, SS generators do not require high voltage to generate microwave energy. Indeed, only a voltage between the 30 and 50 V range is required, which simplifies the equipment design and operators' safety. Another advantage of SS generators is that they are compact and capable for both continuous and pulse operations.

2.2.1.3 Waveguides and applicators

Waveguides propagate, radiate, or transfer the generated energy from the magnetron to the oven cavity. Waveguides are transmission coaxial lines used for guiding electromagnetic waves. At higher frequencies like microwaves, waveguides have lower losses and are therefore used for power applications. Mainly, waveguides are hollow conductors of constant cross section of rectangular and circular forms. Its size defines a minimum frequency fc (the so-called cut-off frequency) below which waves do not propagate. Within the waveguide, the wave may travel in so-called modes, which define the electromagnetic field distribution within the waveguide. These modes can be divided into transversal electric (TE) and transversal magnetic (TM) modes, describing the direction of the electric or the magnetic field, respectively, toward the propagation direction. The most commonly used waveguide is of rectangular cross section with a width equal to double the height in TE10 mode.

The waveguide can be used as an applicator for microwave heating, where the material to be heated is introduced by wall slots and the waveguide is terminated by a matched load. This configuration is then called a traveling wave device, since the location of the field maxima changes with time. A radiation through the slots only occurs if wall current lines are cut and the slots exceed a certain dimension, which can be avoided.

More common in the food industrial and domestic field are standing wave devices, where the microwaves irradiate by slot arrays (that cut wall currents) or horn antennas (specially formed open ends) of waveguides. Three types of applicators depending of the type of field configurations are: (a) near-field applicators, (b) single-mode applicators, and (c) multiple mode applicators.

In the near-field applicators, the microwaves originating from a horn antenna or slot arrays "hit" directly the product to be heated and are almost completely absorbed by it. The transmitted microwaves have to be transformed into heat in dielectric loads (usually water) installed behind the transmitted product. This case is very similar to the traveling wave device, since a standing wave cannot develop. Consequently, no standing wave pattern can be formed, which can yield a relatively homogeneous electrical field distribution (depending on the mode irradiated from the waveguide) within a plane orthogonal to the direction of propagation of the wave. The near-field applicators as well as the traveling wave devices work best with materials with high losses. For substances with low dielectric losses, applicators with resonant modes, which enhance the electric field at certain positions, are better suited. The material to be heated should be located at these positions.

Single-mode applicators consist generally of a feeding waveguide and a relatively small microwave resonator with dimensions in the range of the wavelength. As in the case of the resonator measurements, standing wave (resonance) exists within the cavity at a certain frequency. The standing wave yields a defined electric field pattern, which can then be used to heat the product.

With increasing the dimensions of the cavity, a transition from the single mode to the multimode applicator occurs due to the fast growth of mode density with applicator size and the fact that microwave generators like magnetrons do not emit a single frequency but rather a frequency band. In industrial and domestic applications, the multimode applicators play the most important role, since most of the conveyor-belt-tunnel applicators and the microwave ovens at home are of the multimode type due to their typical dimensions. Despite the high number of stimulated modes, a nonhomogeneous field distribution (constant in time) will develop depending on the cavity and the product geometry and the dielectric properties of the material to be processed. In contrast to the single-mode application, normally this inhomogeneous field distribution, which would result in an inhomogeneous heating pattern is not desired. Possible solutions are either moving the product on conveyor belt or turn tables or changing the field configuration by varying cavity geometries using mode stirrer. Stirrer—the stirrer is usually a fan-shaped distributor which rotates and scatters the transmitted energy throughout the oven. The stirrer disturbs the standing wave patterns, thus allowing better energy distribution in the oven cavity. This is particularly important when heating nonhomogenous materials like foods.

2.2.1.4 Domestic microwave oven

Microwave ovens are broadly used around the world as kitchen appliance since the functions of fast reheating of refrigerated RTE products and defrosting of frozen meals are convenient and efficient. A microwave oven is in nearly every US home—90% of households have one, according to the US Bureau of Labor Statistics. The microwave oven was invented at the end of World War II by Percy LeBaron Spencer. First microwave ovens were too big and expensive, and there was a concern in terms of the radiation they use. By the 2000s, the Americans named the microwave oven as the number one technology.

Microwave reheating of refrigerated or frozen meals can result in equal or slightly better retention of products' quality and nutrients as conventional oven reheating.

The household microwave oven consists of a magnetron tube, which is coupled by a waveguide and an aperture to a rectangular cavity. Due to reflections, a standing wave is developed, which leads to an inhomogeneous heating pattern even in a homogeneous sample. Three possible solutions can effectively reduce this heating behavior. The simplest method is to use low microwave power mostly attained by pulsing the microwave irradiation. The consequence is that the relatively slow heat conduction mechanism equalizes the temperature gradients within the product. Another solution is the use of turntable of microwave ovens. The movement of the dielectric material enhances the power uniformity in two ways: by averaging different electric field strength areas and changing the electric field pattern due to the varied geometrical setup, which yields different field configurations. Concurrent to this turntable, mode stirrers can be used, which are placed just before the aperture of the waveguide to the cavity. This mode stirrer also changes the geometrical setup of the complete cavity and therefore yields time-dependent field configurations, leading to more even product heating. In most cases, these methods have to be combined to achieve necessary relatively uniform heating.

The radiation produced by a microwave oven is nonionizing. Therefore, it doesn't have the cancer risks associated with ionizing radiation such as X-rays and high-energy particles. Long-term rodent studies to assess cancer risk have so far failed to identify any carcinogenicity from 2.45 GHz microwave energy even with chronic exposure levels far larger than humans are likely to encounter from any leaking ovens. The main safety concern found in the use of microwave ovens is that, in general, microwaves heat unevenly and can cause parts of the food to either be undercooked or overcooked. Caution is needed—as well as a few extra minutes—for the heat to equalize within the food. The primary injury that results from using a microwave oven is a burn resulting from hot food and liquids or the particles of hot food from explosions from foods, such as eggs in their shells, cooking unevenly. The use of metal pans made for convectional ovens or aluminum foil is not recommended as they can cause uneven cooking and could even damage the oven. Direct microwave exposure is not generally possible, as microwaves emitted by the source in a microwave oven are confined in the oven by the material out of which the oven is constructed. Furthermore, ovens are equipped with redundant safety interlocks, which remove power from the magnetron if the door is opened. This safety mechanism is required by USA federal regulations.

2.2.1.5 Smart microwave oven
Heating uniformity of foods in microwave oven is very important for reheating and defrosting, which can be partially improved by rotating turntable or mode stirrers. In addition, heating uniformity can be further improved by using sensors or adjusting the power control algorithm or changing the geometry structure.

Samsung has developed a microwave oven that offers a variety of cooking methods. In addition to defrosting meats and reheating leftovers, the oven can fry and bake. It also has a fermentation cycle that can be used in making fresh dough and yogurt.

A microwave oven from NXP Semiconductors uses SS RF energy to cook. The microwave oven controls where, when, and the amount of energy that is transmitted directly into the food. This results in improved consistency, taste, and nutrition of meals. The SS device allows for controlling large amounts of energy with high efficiency and with real-time feedback.

Microwave ovens are also gaining features to be able to connect to mobile technologies, such as the line of LG smart appliances. These appliances have the ability to be turned on and off remotely from anywhere via a smartphone or other device. However, this kind of smart microwave oven is not a real intelligent device since there is no state-sensing and closed-loop control. Closed-loop control for industrial or civil microwave ovens is prevalent, and temperature, humidity sensors are adopted to acquire the feedback signals.

There are mainly two kinds of sensors for microwave ovens—contact sensor and noncontact sensor. The contact sensor is mainly a food probe, such as a negative temperature coefficient (NTC) temperature probe with microwave shielding. However, microwave shielding process is very complex and the reliability needs to be verified. Despite that fiber optic temperature probe can be applied to replace NTC probe to avoid the microwave shielding process.

For noncontact measurement, weight sensor and humidity sensor can be applied for feedback control. The target heating power and time are determined by a fuzzy controller based on the food weight measured by a pressure sensor mounted on the bottom of the microwave oven. The controller is capable to decide the rest heating time to finish once the steam is detected by humidity sensor and the amount reaches the preset threshold.

Additionally, an infrared (IR) fiberoptic radiometer can be installed above the sample to monitor the thermal radiation emitted from the sample. Because IR sensor is expensive, IR array sensor and humidity sensor can be used. These two sensors are complementary, and their fusion processing algorithms are developed to achieve the accurate closed-loop control for food reheating and defrosting purposes. According to the report by He et al. (2022), a truly smart microwave oven can be built using these two sensors, and reheating or defrosting of different food products can be achieved via using one button on a control panel. Automatic control for reheating has been achieved through fuzzy classification of the heated food and energy estimation, while automatic control for defrosting is implemented by temperature-rise control. The authors claimed that the proposed smart microwave oven is low cost and easy to implement.

2.2.1.6 Industrial microwave systems

The use of microwave energy in food processing can be classified into nine unit operations described in the Chapter 1: preheating, baking and precooking, tempering, blanching, frying, extraction, pasteurization and sterilization, and dehydration. Although the objectives of these processes differ, they are established by similar mean to increase in temperature of the product. For each operation, different advantages and disadvantages have to be taken into account.

Industrial applications mostly need continuous processing due to the desired high throughputs. The industrial systems can be divided into two groups by the number and power of microwave sources: high-power single magnetron and low-power multimagnetron devices. Whereas for a single-mode unit, only a single source is possible, in all other systems (multimode, near-field, or traveling wave system), the microwave energy can be irradiated by one high-power magnetron or several low-power magnetrons. An important hurdle for all continuous ovens is the avoiding of leakage radiation through the product inlet and outlet. The leakage radiation is limited to 5 mW cm^{-2} at any accessible place. For fluids or granular products with small dimensions (cm-range), this value can be achieved by inlet and outlet sizes together with the absorption in the entering product or using additional dielectric loads just in front of the openings. In the case of larger product pieces, inlet and outlet gates, which completely close the microwave application device, have to be used.

Recently, cylindrical cross section cavities (applicators) have been applied for industrial pasteurization and sterilization of pumpable food fluids in continuous flow. Otherwise, rectangular cavities have been explored for microwave-assisted tempering, freezing, and also pasteurization of solid foods and prepacked foods.

2.2.2 Heating characteristics of microwave systems

Similar to light, microwaves may be reflected or partly absorbed by the material, may be transmitted through materials without any absorption, e.g., packaging materials and when traveling from one material to another, the microwaves may change direction. The absorbed microwave energy by the product results in its heating. Characterization of the heating efficacy of the microwave processing systems for liquids and solid foods, with or without packaging is essential for the evaluation of system efficiency. This may include determination of a few parameters and selection of the conditions of the best system performance to heat a specific product.

Representative heating characteristics of microwave processing system that can be evaluated include:

1. Incident microwave power, (W or kW)
2. Microwave absorbed power per product volume, (W/m^3)
3. Coupling coefficient
4. Microwave heating rates—volume temperature increases in the product per unit of time, (ΔT, °C/s)
5. Time-temperature heating curves during transient and steady-state conditions (T, °C vs. time)
6. Spatial temperature distribution in the product and microwave heating uniformity (T, °C vs. distance in X, Y, Z-coordinate, m)

1. Incident microwave power (P) in watts is the nominal microwave power generated by the magnetron. The increasing or decreasing of incident power directly affects the heat generated in the sample per unit time, then the heating rate of the sample is affected as well. In reality, microwave energy is partially absorbed by the load, and another part (i.e., reflected microwave energy) propagates forth and back within the applicator. In domestic microwave ovens, incident microwave power can be simply evaluated by measuring the time to heat 1 kg of pure water at 20°C in a glass container up to 100°C. Additionally, energy consumption is an important measurement for a productive process, so the total energy of different input power consumed before the phase transition temperature should be considered to choose an optimal input power.

2. The absorbed power for the product load (P_{abs}, W) can be calculated according to Eq. (2.1):

$$P_{abs} = mC_p \Delta T \qquad (2.1)$$

where m is a product mass (kg) or flow rate in kg/s; Cp is a specific heat capacity (J/kg °C); and ΔT is a temperature difference between the final T_f and initial T_0 temperatures of the product.

The dielectric properties and specific heat capacity, Cp, play an important role how the products will be heated by microwave energy. Cp is the amount of heat (in Joules) needed to raise the temperature of 1 kg of matter by one degree at a given temperature. Specific heat capacity of solids and liquids depends upon product composition and temperature but does generally not exhibit pronounced pressure dependence. It is common to use the constant pressure specific heat, Cp, which thermodynamically represents the change in enthalpy H (kJ/Kg) for a given change in temperature T when it occurs at constant pressure P.

3. Coupling coefficient is calculated as a ratio of absorbed energy (P_{abs}) and heat losses to nominal incident power (P) of the microwave heating system and typically is less than 1. Coupling coefficient should be checked for various levels of input power, product volumes, and shapes. Some authors call this ratio of absorbed power and input power as absorption efficiency. It is used to describe the microwave energy absorption capability of the product.

4. Microwave heating rates are evaluated as average volume temperature increase (ΔT, final product temperature minus initial temperature) per unit of heating time (°C/s). Microwave heating rates are compared for different levels of the input power. In continuous flow microwave systems, in order to calculate microwave heating rates, volume averages liquid product temperature rise is used versus flow rate or average residence time or treatment time (°C/s) and compared for different levels of the input power.

5. Time-temperature heating curves are measured during transient and steady-state heating conditions in multiple locations of the product volume near the center and wall.

6. Knowledge of spatial temperature distribution is critical for the determination of a location of the least heated point in the product and further calculation of process lethality in this point. In microwave heating, the location of the least heating point with minimum lethality often doesn't coincide with the geometrical center of the product. Microwave heating uniformity is particularly important in systems intended for preservation purposes. A homogeneous temperature distribution is crucial to ensure that microbiological requirements are fulfilled, without unnecessary overheating of the food. The lack of uniformity of heating has been a significant technical hurdle for the application of microwave energy. Achieving product heating uniformity requires an appropriate design of the microwave system, including wave guide system, applicators, and cavities. Modeling tools can be used to predict the location of "cold spots" for appropriate designs and configurations of the system.

2.2.2.1 Flow regime
In a case of continuous flow microwave process of liquid products, juices, and beverages, temperature rise versus flow rate, residence time, and at different levels of input power need to be known. Additionally, the type of flow regime (turbulent or laminar) and mixing efficiency should be known through the calculation of dimensionless indicators such as Reynolds (Re) and/or Dean (De) numbers because flow conditions can greatly affect the product heating uniformity.

2.2.3 Effect of product geometry
Geometric shape of the product, packaging, and heating chamber can influence the reflection of microwave electromagnetic energy, power absorption efficiency, and consequently the depth of penetration into the product from a variety of directions. Also, the position of the product in the heating chamber is critical for heating uniformity. The general understanding is that a spherical shape of the product seems the ideal model because a sphere dimension is uniform in all directions. Other geometric shapes may cause nonuniform heating due to edge effects.

In terms of the time required to heat the product to the desired temperature, smaller dimensions in thickness and product volume attenuate energy more readily in relatively small or thin samples. For large volumes, longer time is required to reach the target temperature and therefore more time to diffuse the energy thereby reducing the uniformity of temperature distribution. Using agar as an example of heating medium, it was shown that the time needed to heat the plate with agar is shorter than other shapes because the size of a plate of agar is smaller than the size of a sphere or cylinder. The coldest point in each model was located in a central position of sample (Soto-Reyes et al., 2015).

Zhang et al. (2018) applied COMSOL Multiphysics software to model microwave heating (temperature distribution and power absorption) of potato samples of various geometrical shapes and dimensions. It was demonstrated that in the microwave heating cavity, samples with equal volume and different shapes form hot spots at different positions in the center or near the surface of samples. Choosing proper input power can achieve the goal of high temperature distribution uniformity and lower energy consumption. The samples of spherical and cubic shapes can be classified into four groups according to the microwave power absorption as well as the locations of electromagnetic focusing and hot spots. For the small and big volume groups samples, spherical samples are better than the cubic ones. For the intermediate size samples, if temperature uniformity of samples is more important, spherical samples are the best choice, while if the decisive factor is the high microwave power absorption capability, cubic samples have been recommended. As for the isometric volume cylindrical, cuboidal, and ellipsoidal samples, they can be classified into three groups by the wave numbers and penetration numbers in the radial and axial directions. In accordance with volume or the wave numbers and penetration numbers in the radial and axial directions of samples, they can be classified into different groups.

Therefore, when considering the appropriate volume and shape of the sample, their effects on microwave system parameters should be considered as well.

When the distribution of temperature is not important, shapes with edges and corners are favored, and when uniform heating is needed, spherical shapes are preferable. Numerical simulation is an essential tool to predict how the shapes, dimensions, or cross section of food materials (squares, circular, etc.) may play a role in microwave heating. The square or rectangle shape that exhibits corner heating patterns and may not be recommended food shapes for efficient industrial microwave processing.

2.3 Dielectric properties of foods

In addition to the system design, to a large extent, microwave heating characteristics of foods are determined by dielectric and physical properties of the treated product, its composition, and variations in the composition. The dielectric properties that are relevant in microwave heating are permeability, permittivity (capacitivity), and electrical conductivity of the heated material.

The permittivity which determines the dielectric constant, the dielectric loss factor, and the loss angle influences the microwave heating. The dielectric permittivity is a complex number used to explain interactions of foods with electric fields. It determines the interaction of electromagnetic waves with matter and defines the charge density under an electric field.

The real part of the complex permittivity, ε', is often called the dielectric constant or "capacitivity" The dielectric constant is a measure of the capacity of a material to store electric energy and a measure of the induced dispersion in a material. It is a constant for a material at a given frequency. A significant change in dielectric properties over a frequency range is called a dielectric dispersion. The relative permittivity is a measure of the polarizing effect from an external field, that is, how easily the medium is polarized.

The imaginary part of the complex permittivity is the dielectric loss factor, ε'', which measures the dissipation of electric energy into heat. A material with high values of the dielectric loss factor will absorb energy at a faster rate than materials with lower loss factors, since electrical power transferred to the unit volume of load as heat is directly proportional to the loss factor as well as the frequency and the square of electric field inside the load. The tangent of dielectric loss angle (tan δ) is called the loss tangent or the dissipation (power) factor of the material. This is equivalent to the ratio (Eq. 2.2) of the dielectric loss factor (ε'') and dielectric constant (ε').

$$\tan \delta = \varepsilon''/\varepsilon'' \tag{2.2}$$

In practice, the dielectric properties of materials relative to that of vacuum are used, that is, the relative complex permittivity ($\varepsilon_r^* = \varepsilon^*/\varepsilon_0$) and the relative loss factor ($\varepsilon_r'' = \varepsilon''/\varepsilon_0$). ε_0 is the permittivity in a vacuum, which is equal to 8.85419×10^{-12} F/m. The relative complex permittivity (ε_r^*), relative permittivity

(ε_r), the relative loss factor (ε_r''), and the loss tangent (tan δ) are related as follows (3):

$$\varepsilon_r = \varepsilon/\varepsilon_o = \varepsilon''\varepsilon_o - j\varepsilon''/\varepsilon_o = \varepsilon_r'' - j\varepsilon_r'' = \varepsilon_r''(1 - \tan\delta) \quad (2.3)$$

where ε is permittivity (F/m). Therefore, Eq. (2.2) can also be written as:

$$\tan\delta = \varepsilon_r''/\varepsilon_r'' \quad (2.4)$$

Conductivity (σ) can be determined for a material when ε'' is known as in Eq. (2.5).

$$\sigma = \omega\varepsilon'' = \omega\varepsilon_o\varepsilon_r' \quad (2.5)$$

where σ is in S/m and ω (rad/s) is angular velocity which is equivalent to $2\pi f$.

2.3.1 Effect of foods' dielectric properties on microwave heating

Both dielectric properties are dimensionless because they are relative values. In foods, permittivity can be related to chemical composition, physical structure, frequency, and temperature, with moisture content being the dominant factor.

The effect of permittivity on temperature increase is represented in the following Eq. (2.6) (Orfeuil, 1987):

$$\Delta T = \frac{2\pi t f \varepsilon_o \varepsilon_r \tan\delta V^2}{C_p \rho} \quad (2.6)$$

where ΔT, temperature increase (°C); t, time of temperature rise (s); ε_o, dielectric constant of a vacuum, considered equal to 8.85419×10^{-12} F/m; f, frequency; ε_r', relative dielectric constant or permittivity of the material to be heated; tan δ, tangent of dielectric loss angle; V, electric field strength, equal to voltage/distance between plates, (V/cm); C_p, specific heat of the material to be heated (J/kg/°C); and ρ, density of the material to be heated (kg/m³).

As Eq. (2.5) shows, ΔT can be increased by increasing the loss factor. However, if the loss factor is too high, current leakage takes place through the material. On the other hand, if the loss factor is too low, heating takes place slowly and it becomes difficult to reach the desired temperature due to heat losses. Therefore, for dielectric heating to be successful, the loss factor should lie between $0.01 < \varepsilon'' < 1$. If the loss factor increases with temperature, power can concentrate at the hottest points, which may cause local overheating. In such situations, the power density must be reduced.

Materials which show no dielectric heating are those which contain no free ions and no unsymmetrical molecules (e.g., polystyrene) or those with bound ions in a lattice structure (i.e., low loss factor). Most readily heated materials contain unsymmetrical molecules such as polar molecules and/or free ions or conducting components. Water, which has one dipolar molecule in the absence of any other electric field, absorbs the energy of a microwave electric field very easily.

2.3.2 Effect of food electrical conductivity on microwave heating

The conductivity and the loss factor are closely related (Eq. 2.5), and the influence of the conductivity is usually lumped in with the effect of the loss factor. Electrical conductivity is a measure of how well electric current flows through a material with a certain cross-sectional area A, a length L, and a resistance R. It is the inverse value of electrical resistivity (measure of resistance to electric flow) and is expressed in SI units of S/m in the following relation. The electrical conductivity increased with temperature and salt concentration. Salt ions act as conductive charge carriers. They simultaneously depress the permittivity and increase the absolute dielectric loss, when compared with the behavior of pure water. Depression of dielectric constant (ε') results from the binding of free water molecules by counterions of dissolved salts, and the elevation of dielectric loss factor (ε'') results from the addition of conductive charge carriers (dissolved salts). Both ε'' and ε'' depend on the concentration of the aqueous ionic solution, as well as on the frequency f and the temperature. The dielectric loss of an aqueous ionic solution may be expressed with an equation which represents the combined effects of dipole rotation (D) and conductive (ionic) charge migration (Eq. 2.7).

$$\varepsilon'' = \varepsilon''_D + \varepsilon''_\sigma \tag{2.7}$$

These effects vary inversely with temperature. The dipole loss component ε''_D decreases and the ionic loss component ε_σ'' increases with increasing temperature. Usually, a high ionic conductivity is accompanied by large values of the loss factor. However, if the conductivity of the heated material is too high, a low voltage must be used to prevent current leakage through the product. As seen in Eq. (2.6), this would result in a reduction of heating.

2.3.3 Effect of other physical and thermal properties on microwave heating

The microwave power absorption over certain bands of frequency is characteristic of the structure of molecular or macroscopic scale and is not necessarily associated with their conductivity. The specific heat does not influence the dielectric properties, but it certainly affects the resulting ΔT as seen in Eq. (2.6) where it is inversely proportional to the ΔT. A material with greater specific heat will undergo a smaller temperature change since more energy is required to increase the temperature of 1 g of the material by 1°C. For example, fats and oils, although they are good conductors of heat, exhibit relatively low dielectric properties and heat quickly in dielectric field because of their low heat capacities.

The product density is also inversely proportional to the ΔT. However, it can influence the electrical properties. The applied voltage and the time of heating affect the temperature change in a more obvious way. This follows the theory presented in Eq. (2.6). An increased heating time, which allows more friction to occur in a given space, will also increase the ΔT.

2.4 Factors influencing dielectric properties of foods

The dielectric properties of food products depend mainly on frequency, temperature, water content, chemical composition, and structure. Molecules, which have permanent dipole moments, must be considered when looking at the effects of frequency.

2.4.1 Temperature

As temperature increases, the relaxation time decreases and the loss factor peak will shift to higher frequencies. In a region of dispersion, the loss factor will either increase or decrease depending on whether the operating frequency is higher or lower than the relaxation frequency. In a region of polar dispersion, the dielectric constant will increase with increasing temperature, while outside such a region, it will decrease with temperature (Nelson, 1990). Values for ε_r'', ε_r'', and their temperature dependence for mixtures with high water content indicate that both ε_r' and ε_r'' decrease with increasing temperature at 3 GHz; however, ε_r'' decreases more rapidly unless electric conductivity is high (De Loor, 1968). Dielectric properties at high temperatures useful for microwave pasteurization and sterilization have been scarce. In addition to temperature effects per se., physical and chemical changes such as gelatinization of starch and denaturation of protein at higher temperatures can significantly change dielectric properties. Such effects are also strongly dependent on moisture content. Many studies of dielectric properties were reported in the literature, but the data were not obtained above the boiling point except for Mudget (1986) and Wang (2003) that is critical for microwave sterilization.

2.4.2 Moisture content

There is a strong positive correlation between ε_r' and water content, while the correlation for ε_r'' is uncertain (Bengtsson and Risman, 1971). Measurements by Van Dyke et al. (1969) showed that ε_r'' was nearly independent of water content below 20% and above 45%, while in between, there was a linear relationship with a sharp increase in ε_r'' with moisture content. In food materials, water may be associated with other components forming layers of water, which affects its structure and properties.

Chemically bound water exerts less influence on the dielectric properties of the material than nonassociated water. Individual curves displaying the variation of the dielectric constant or dielectric loss factor plotted against moisture content typically have two distinct slopes. There is a slight increase at low moisture levels followed by a sharp increase when the moisture content reached a certain level. Padua (1993) who showed that sugar solutions at lower concentration (0—0.6 g/g water) consist of units of sugar molecules with eight attached pairs of water molecules rotating freely in the continuous water medium developed the theory of two hydration stages for sucrose-filled agar. The increase in dielectric loss factor for sucrose indicates that sucrose molecules stabilized the hydrogen-bonded structure of water and dissipates electromagnetic energy more effectively. According to the author, concentrations from 0.6 to 1.2 g sucrose/g water, sucrose molecules associate through water bridges forming loosely held clusters. The decrease in loss factor suggests that as the amount of sucrose increases, water molecules participate in more than one hydrogen bond and hinders their rotational movement. Otto and Chew (1992) also found that the dielectric constant is a function of the water-to-solids ratio for Portland cement pastes and mortars, but the magnitude of the electric conductivity (proportional to the loss factor) responds to the type of additive mixed with the cement.

Roebuck and Goldblith (1972) similarly found that the dielectric constant of sugar solutions and starch—water mixtures decreased with water content, while the loss factor changed significantly with the addition of sugars and gelatinized potato starch. The dielectric loss of starch remained constant indicating that a small portion of the water molecules tightly held by starch decreased the mobility, but a large proportion of the water remained free. This indicates the importance of not only the water content but also the chemical composition of the material.

2.4.3 Chemical composition

It is generally known that adding salt increases the dielectric value, but it is not known to what extent. Li and Barringer (1997) found that although a larger salt content in ham samples resulted in a larger dielectric loss increase, the dielectric constant (ε') was relatively constant for different salt contents (0.5%, 1.0%, 1.7%, 2.6%, and 3.5%). This means that a change in salt affects the power factor more significantly than the dielectric constant.

Bengtsson and Risman (1971) found that when 1% salt was added to a gravy, ε_r' changed very little, while ε_r'' increased by about 20%. The added salt raised the ε_r'' values slightly, while the ε_r'' values increased to a much higher extent because of the contribution from electric conductivity to the effective loss factor. Little et al. (1968) reported a wide range of electrical conductivity within a given product. For example, replicate measurements of 25 milk samples varied in conductivity from 4.06 to 6.29 mS/cm at 25°C. The differences were assumed to be due to the variability of leukocyte counts, as they found a correlation between conductivity and leukocyte counts. At 25°C, ε_r'' for reconstituted freeze-dried beef was reported to be nearly independent of water content above 45%, while the value of ε'' more than doubled from 17 to 46 F/m for an increase in salt from 0.6% to 3.8%. The addition of 4% salt to pork doubled ε_r'' and 1% salt increased ε_r'' by about 20%.

Salt addition can also decrease the dielectric constant as reported by Hasted et al. (1948). Piyasena and Dussault (1999) observed that the effect of salt content was not significant within the concentrations tested between 0.2% and 0.70% in 0.27% guar solution. The dielectric constant may be lowered by a concentrated salt solution because of the saturation of the dielectric in the neighborhood of an ion.

2.4.4 Effects of nonelectrolytes in water

Sugar is a key dielectric component in food. The importance of sugar as a microwave-absorbing food ingredient may supersede its conventional role as a sweetener, plasticizer, and texturing agent and as a water activity control agent. Addition of sugar increases the boiling point and changes the dielectric properties of sugar–water mixtures. Solutes such as sugars modify the dielectric behavior of water. Solutions containing mixtures of sugars are often twice as absorptive as the individual components of the binary mixture. According to the reference (Shukia and Anantheswaran, 2001), this interactive behavior offers a definite advantage in microwave cooking of sugar-based food formulations. Aqueous solutions of sugars and alcohols are unusual in their dielectric behavior. The loss factor profile with respect to the percent of sugar or glycerol in water exhibits a maximum at 50%.

2.4.5 Effect of pH and ionic strength

Hydrogen ion concentration of a food determines the degree of ionization. Dissociated ions migrate under the influence of an external field, and hence the pH becomes a factor in microwave heating. Ionic strength determines the collision frequency and the heat generation. Collision increases at higher concentration of ions to a point that dielectric loss factor may have a positive temperature effect.

2.4.6 Organic solids

The dielectric properties of structural, suspended, or bound food constituents classified in proximate analysis as moisture, carbohydrate, lipid protein, or ash content are similar to those for a variety of inorganic and organic solids that are dielectrically inert compared with dielectrically active fluids like water or aqueous ions. The major effect of undissolved structural or colloidal solids on food behavior is to depress levels of dielectric activity by excluding dielectrically active materials from the total volume occupied by the product. The properties of undissolved food solids are similar to those of ice at temperatures near the freezing point and are relatively independent of frequency and temperature. Similar behavior and levels of dielectric activity have been observed in measurements of many inorganic and organic solids, fats, and oils.

2.4.7 Proteins

Solid food proteins are dielectrically inert. However, both dielectric constant and dielectric loss increase with an increasing amount of water in a range of 2%–12%. The hydrated form of protein, protein hydrolysates, and polypeptides are much more microwave reactive. Free amino acids are also very reactive and contribute to increase in dielectric loss. A considerable difference in microwave reactivity is expected in cases of cereal, legume, milk, and meat proteins. Knowledge of dielectric behavior and properties of hydrated proteins is of critical importance in the design of microwave heating process.

2.5 Propagation of microwave electromagnetic waves

Microwave energy considered as a radiation when microwaves impinge on a surface, some of it will be reflected, some of it will be refracted and transmitted and partially be absorbed, while generating heat. Most of the remaining energy will be reflected back at other food surfaces and so on. Depending on the food geometry, the result can be the focusing of energy on certain areas and may be part of the explanation for selective heating effects or energy concentration. For spherical or cylindrical shapes, the result can be a concentration of energy to the center of food, depending both on the food dimensions and dielectric properties. This central heating effect occurs for diameters approximately one to three times the penetration depth in the material. For cylinders, concentration effects occur when the electrical field is parallel to the cylindrical axis. The effect is stronger for foods with high values of its dielectric properties. At 240 MHz, central heating effect usually happens at diameters between 25 and 55 mm, while the values are correspondingly larger for 915 MHz (Ohlsson and Risman, 1978).

Materials can be classified into reflecting, absorbing, and transparent according to their interaction with the electromagnetic field. Reflecting materials are mostly metals, where electromagnetic field creates surface currents and only penetrate a few microns into material. Transparent materials do not absorb the energy to any significant extent. Glass and most plastics are typical examples. Absorbing materials are those containing polar constituents, predominantly water will absorb microwave energy and heated accordingly to the mechanism explained above. For heating, we would like to maximize absorption. We would like packaging to be transparent and our shielding to be reflective.

2.5.1 Transmission properties of foods

The distribution of electrical energy coupled from an electromagnetic field by radiative transfer is determined by asset of transmission properties related to a material's dielectric properties and the material's thermal properties (thermal conductivity, density, heat capacity, i.e., thermal diffusivity). The material's transmission properties may be defined in terms of a complex propagation factor (Mudgett, 1986):

$$\gamma = \alpha + j\beta \tag{2.8}$$

Involving attenuation and phase shift of a traveling wave as it is propagated in a dielectric material. Attenuation factor α is related to the material ability to attenuate or absorb electrical energy coupled by material from EM field and is the reciprocal of the material's penetration depth $d = 1/\alpha$.

Bengtsson and Risman (1971) defined the penetration depth as the depth in a material where the energy of a plane wave propagating perpendicular to the surface has decreased to $1/e$ of the surface value:

$$d = \frac{c}{2f\pi\sqrt{2}\sqrt{\varepsilon'_r}\sqrt{\sqrt{1 + (\tan\delta)^2} - 1}} \tag{2.9}$$

where c is the speed of propagation of waves in a vacuum, which is equal to 3×10^8 m/s and d is in meter. In most cases, tan δ is low (i.e., far less than 1) and Eq. (2.8) can be simplified (or rewritten) as Eq. (2.10):

$$d = \frac{c}{2\pi f \sqrt{\varepsilon'_r} \tan \delta} = \frac{4.77 \times 10^7}{f \sqrt{\varepsilon'_r} \tan \delta} \quad (2.10)$$

The above Eq. (2.10) illustrates the effect of frequency, dielectric constant, and dielectric loss factor on penetration depth. Bengtsson and Risman (1971) found that the deepest penetration depth was experienced when both ε'_r and ε''_r were low. This is consistent with the theory, which shows that the dielectric loss factor is inversely proportional to the penetration depth (Eq. 2.9). However, the dielectric constant has a greater impact on penetration depth than the power factor (tan δ). If the power factor is much less than 1, its influence will be minimal, as seen in Eq. (2.9). The composition of the material also affects the microwave penetration depth. Bengtsson and Risman (1971) reported a decrease in penetration depth with salt addition and an increase with fat content and decreasing water content. Water was found to have a larger penetration depth than all "wet" food materials such as ham, gravy, mashed potato, broth, carrot soup, peas, beef, cod, and liver at 20°C.

The imaginary component of the propagation factor (eq. 2.8), the phase factor, is a measure of the material's ability to store electrical energy from the applied field and is generally related to a space period. Transmission properties characterize the ability of material to couple EM field power and are also related to the material's dielectric properties.

2.5.2 Wave impedance and power reflection

The reflection phenomena can be analyzed in terms of characteristic wave impedance. Impedance is the ratio between the electric and magnetic field strength according to Eq. (2.11):

$$\eta = \eta_0 / \sqrt{\varepsilon} \quad (2.11)$$

where η_0 is wave impedance of free space equal to approximately 377 Ω. The reflection and transmission at a plane boundary are primarily related to square root of permittivity, and the main determining factor for the magnitude of reflection is from real permittivity of the material. The characteristic impedance for the average food is about 50 Ω. The change in characteristics impedance (the dielectric mismatch) at the food surface results in reflection of about 50% of the microwave power falling on the surface. Most of the energy is reflected back to the food via the metal cavity walls.

The efficiency of energy transfer between the processing equipment and the product may also be affected by electrical properties related to the product's basic dielectric properties. Such effects appear to be related to the nature of the product and load volume as shown in Fig. 2.1 for samples of oil and water heated in a domestic microwave oven at 2450 MHz. As shown, energy transfer efficiencies for

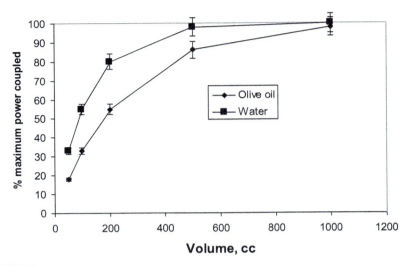

FIGURE 2.1

Energy transfer efficiencies for olive oil and water in a 750 W microwave oven at 2450 MHz*.

*Adapted from Mudgett, R. 1986. Microwave properties and heating characteristics of Foods. Food Technology, 84–93.

water were higher than those for oil at corresponding load volumes of similar geometry. The observed effects suggest that energy transfer efficiencies maybe optimized in microwave food processing by loading with enough product volume. The key relationship that describes impedance-matching effects is Eq. (2.10). Calorimetric measurements for aqueous ions and selected organic solids were similar for water and oil. In practice, loading effects on power coupling may be monitored by voltage standing wave ratios for radiative energy transfer or by direct measurements of power for conductive energy transfer.

Dielectric and physical properties of liquid foods, beverages, and packaging materials are summarized in Table 2.1.

2.5.3 Dielectric properties of mixtures

The permittivity of aqueous solutions or mixtures is decreased by two mechanisms: the replacement of water by a substance with a lower permittivity and the binding of water molecules. When the size of inhomogeneities (particles, grains) is much smaller than the wavelength, the effective permittivity of the mixture depends only on the shape of the inhomogeneities not on their size. Dielectric mixture equations can be used to estimate the dielectric properties of mixtures. The suitable mixture relationships were reviewed by Ryynänen (1995).

Table 2.1 Dielectric and thermal properties of liquid foods, solutions and packaging materials at 2450 MHz.

Food or material	Dielectric constant e'	Loss factor e''	Penetration depth, d_p, cm	Density, ρ kg/m³	Heat capacity C_p (J/kgK)	Thermal conductivity k (W/mk)	Thermal diffusivity (m²/s) × 10⁻⁷
Water and water solutions							
Distilled water	77.4	9.2	1.7	1000	4180	0.6	1.4
Water + 1% NaCl	77.1	23.6	0.73	1005	4100		
Water + 5% NaCl	67.5	71.1	0.25				
Water + 10% sucrose	72.7	10.92		1034	3725		11.7
Water + 30% sucrose	62.7	14.6					
Water + 25% sorbitol	68.4	14.2					
Liquid foods							
Cooking oil	2.5	0.1	23.7	910	2010	0.17	0.9
Milk 1%	70.6	17.6					
Milk 3.25%	68.0	17.6					
Red wine	66.6	19.8					
Red grape juice	65.4	19.9					
Liquid whey protein 915 MHz	61.65	33.55	1.26				

Continued

Table 2.1 Dielectric and thermal properties of liquid foods, solutions and packaging materials at 2450 MHz.—cont'd

Food or material	Dielectric constant e'	Loss factor e''	Penetration depth, d_p, cm	Density, ρ kg/m^3	Heat capacity Cp (J/kgK)	Thermal conductivity k (W/mk)	Thermal diffusivity (m^2/s) × 10^{-7}
Materials							
Soda lime glass	6	0.1	40	2600	670	0.88	5.1
Teflon	2.1	0.0006	4920	2100		0.2	
Nylon	2.4	0.02	1650				
Paper	3–4	0.05–0.111	50				

Sakai et al. (2005) conducted a study of model systems for use in microwave heating when heating pattern could be adjusted and tuned. Agar gel was used as the base in the model foods, and sucrose and sodium chloride were used to adjust the dielectric properties. It was found that the addition of sucrose primarily changed the dielectric constant, whereas the addition of sodium chloride changed the dielectric loss factor. They proposed a procedure for determining the concentrations of sucrose and sodium chloride to be added when applying a certain dielectric constant and loss factor. The experiments were conducted to mimic a gratin sauce and agreement between temperature increases in the model system, and this was found in real food.

2.6 Dielectric properties of foods in radio frequency range

The dielectric properties that are relevant in RF heating are similar to dielectric properties in microwave frequency range: permeability, permittivity (capacitivity), and electrical conductivity of the heated material. The permittivity which determines the dielectric constant, the dielectric loss factor, and the loss angle influences the efficiency of RF heating. A product with high values of the dielectric loss factor absorbs energy at a faster rate than materials with lower loss factors because electrical power transferred to the unit volume of load as heat is directly proportional to the loss factor as well as the frequency and the square of electric field inside the load (Piyasena et al., 2003). However, a significant change in dielectric properties of materials and products can be observed over a frequency range that is called a dielectric dispersion. The power absorption over certain bands of frequency is characteristic of the structure of molecular or macroscopic scale and is not necessarily associated with the conductivity. The dielectric constant of materials either decreases or remains constant as frequency increases. The loss tangent and the dielectric loss factor may either increase or decrease depending on the starting relaxation time. Additionally, it was reported that the energy efficiency and heating rate will be maximized at or near the location in frequency of a "Debye resonance"—the frequency where the dielectric loss factor is at a maximum for a particular material. Multiple Debye resonance might occur in the complex products.

Similar to dielectric properties in microwave range, other factors that impact dielectric properties in RF range include temperature, moisture content, and chemical composition. The temperature and frequency dependence are closely related in theories of dielectric dispersion as detailed in the work by Ohlsson et al. (1974).

The dielectric properties of food materials in the RF and microwave region can be determined by several methods using different measuring devices. Methods of measuring these dielectric properties vary even in a given frequency range. Ryynanen (1995) reviewed in details four groups of measurement methods: lumped circuit, resonator, transmission line, and free space methods. The suitable method can be selected based on the samples size, temperature range, and frequency. The amount of available information on the dielectric properties of foods in the RF range is limited in comparison with considerable amount of data on dielectric properties of foods at microwave frequencies. Piyasena et al. (2003) presented a comprehensive data of dielectric properties of selected foods in the RF frequency range 1−200 MHz including meat, poultry, and fish (raw, frozen, and cooked), agricultural products (rice, wheat, corn, nuts, peas), fresh produce (fruits and vegetables), bakery products (raw and baked dough, tortillas), and biological tissues. The RF properties of some foods (for example, dairy liquids, vegetables, fruits, and various foods components) could not be found in the literature and should be measured or estimated as functions of frequency, temperature, moisture content, or other processing parameter.

2.7 Conclusions

In summary, the determination of microwave heating characteristics such as incident microwave power, microwave absorbed power per product volume, coupling coefficient, product heating rates, time-temperature heating curves during transient and steady-state conditions, and spatial temperature distribution in the product are a few parameters that are essential for the selection of the optimal conditions of best system performance to heat a specific product and achieve microwave heating uniformity. Additionally, geometrical dimensions, shape, and position in the microwave chamber are essential for process uniformity and higher efficiency.

Dielectric properties provide an indication of the electrical insulating ability of the material. Foods are poor insulators; therefore, they generally absorb a large fraction of the energy when placed in a microwave field resulting in instantaneous heating. During microwave processing, dielectric properties and heat capacity play essential role in coupling incident microwave power and its transfer into the treated material.

The chemical composition of food with respect to water, salts and ions, and sugar is the major determinant of their dielectric properties. Addition of proteins, starch, and gums can further modify these properties. Collectively these ingredients in a formula can be varied to control the rate of heating, depth of heating, and efficiency of microwave coupling.

The incident microwave power, heat capacity, and density of heated material are parameters that effect product temperatures and heating rates. Depth of penetration and thermal conductivity of heated material influence the shape of temperature profiles and the temperature spread in the layer (evenness of temperature distribution).

Also, dielectric properties are included in the governing conservation equations and dependent on temperature, the material composition, and the selected frequency, but not affected by the electromagnetic field strength.

According to the literature, the important factors, which influence the dielectric properties of heterogeneous mixtures containing water such as majority of foods will be water content and the way in which the water bound to the material and the salt content or electric conductivity.

Similar to microwave heating, the effectiveness of RF heating is highly dependent on the products' dielectric properties. Dielectric properties of food products as functions of temperature or variation in composition are essential part in successful development of RF-based operations in numerous food applications. Therefore, the parameters affecting dielectric properties of foods should be taken into account when designing an RF-based heating system for foods.

2.8 Nomenclature

c	speed of propagation of wave in vacuum (3×10^8 m/s)
Cp	specific heat of the material to be heated (J/kg-°C)
D	electrolyte thickness (μm)
f	frequency (Hz)
T	membrane thickness (nm)
t	temperature rise time (s)
V	electric field, equal to voltage/distance between plates, (V/cm)

Greek Letter and Symbols

T	temperature increase (°C)
δ	dielectric loss angle (°)
ε	permittivity or capacitivity of the material (F/m)
ε'	real part of complex permittivity or dielectric constant of the material (F/m)
ε''	imaginary part of complex permittivity or dielectric loss factor of the material (F/m)
ε_∞	relative dielectric constant at high frequency
ρ	density of the material to be heated (kg/m³)
σ	conductivity associated with conduction current in the material (S/m)
τ	relaxation time (s)
ω	angular velocity (rad/s)
σ_c	complex conductivity associated with conduction current in the material (S/m)
$\sigma_{c'}$	real part of conductivity associated with conduction current in the material (S/m)
ε'_r	static or DC value of relative dielectric constant
ε''	ionic loss component (F/m)
$\sigma_{c''}$	imaginary part of conductivity associated with conduction current in the material (S/m)
ε_D''	dipole loss component (F/m)
ε_o	dielectric constant of vacuum (8.85419×10^{-12} F/m)
ε_r	relative permittivity or capacitivity ($\varepsilon/\varepsilon_o$)
ε_r''	relative dielectric loss factor ($\varepsilon''/\varepsilon_o$)
ε'_r	relative dielectric constant of the material ($\varepsilon'/\varepsilon_o$)

References

Bengtsson, N.E., Risman, P.O.(, 1971. Dielectric properties of foods at 3 GHz as determined by cavity perturbation technique. Journal of Microwave Power 6 (2), 107–123.

De Loor, G.P., 1968. Dielectric properties of heterogeneous mixtures containing water. Journal of Microwave Power & Electromagnetic Energy 3 (2), 67–73.

Hasted, J.B., Ritson, D.M., Collie, C.H., 1948. Dielectric properties of aqueous ionic solutions. Parts 1 and 2. Journal of Chemical Physics 16 (1), 1–21.

He, C., Zhong, G., Wu, H., Cheng, L., Huang, Q., 2022. A smart reheating and defrosting microwave oven based on infrared temperature sensor and humidity sensor. Innovative Food Science & Emerging Technologies 77 (2022), 102976. https://doi.org/10.1016/j.ifset.2022.102976.

Li, A., Barringer, S.A., 1997. The effect of salt on the dielectric properties of ham at sterilization temperatures. IFT Annual Meeting Book of Abstracts 55–5, 155.

Little, T.W.A., Hebert, C.N., Forbes, Z., 1968. Electrical conductivity and the leucocyte count of bovine milk. The Veterinary Record 82, 431–433.

Mudgett, R., 1986. Microwave properties and heating characteristics of Foods. Food Technology 84–93.

Nelson, S.O., Kraszewski, A.W.(, 1990. Dielectric properties of materials and measurement techniques. Drying Technology 8 (5), 1123–1142.

Ohlsson, T., Risman, P., 1978. Temperature distributions of microwave heating spheres and cylinders. Journal of Microwave Power 13 (4), 303–310.

Ohlsson, T., Bengtsson, N.E., Risman, P.O., 1974. The frequency and temperature dependence of di-electric food data as determined by a cavity perturbation technique. Journal of Microwave Power & Electromagnetic Energy 9 (2), 129–145.

Orfeuil, M., 1987. Electric Process Heating: Technologies/equipment/applications. Battelle Press, Columbus, OH.

Otto, G.P., Chew, W.C., 1992. Electromagnetic properties of large-grain materials measured with large coaxial sensors. In: Mc Gonnagle, W. (Ed.), International Advances in Nondestructive Testing, vol 17. Gorden and Breach Science Publishers, New York, NY.

Padua, G.W., 1993. Proton NMR and dielectric measurements on sucrose filled agar gels and starch pastes. Journal of Food Science 58 (3), 603–604,626.

Piyasena, P., Dussault, C., 1999. Evaluation of a 1.5 kW radio-frequency heater for its potential use in a High Temperature Short Time (HTST) process. In: CIFST Annual Conference, Kelowna, BC, June 1999.

Piyasena, P., Dussault, C., Koutchma, T., Ramaswamy, H., Awuah, G., 2003. Radio frequency heating of foods: principles, applications and related properties—a review. Critical Reviews in Food Science and Nutrition 43 (6), 587–606. https://doi.org/10.1080/10408690390251129. To link to this article:

Roebuck, B.D., Goldblith, S.A., 1972. Dielectric properties of carbohydrate-water mixtures at microwave frequencies. Journal of Food Science 37, 199–204.

Ryynänen, S., 1995. The electromagnetic properties of food materials: a review of the basic principles. Journal of Food Engineering 26, 409–429.

Sakai, N., Mao, W., Koshima, Y., Watanabe, M., 2005. A method for developing model food system in microwave heating studies. Journal of Food Engineering 66, 525–531.

Shukla, T.P., Anantheswaran, R.C., 2001. Ingredient interactions and product development for microwave heating. In: Handbook of Microwave Technology for Food Applications. CRC Press, p. 40.

Soto-Reyes, N., Temis-Perez, A.L., Lopez-Malo, A., Rojas-Laguna, R., Sosa- Morales, M.E., 2015. Effects of shape and size of agar gels on heating uniformity during pulsed microwave treatment. Journal of Food Science 80 (5), E1021–E1025.

van Dyke, D., Wang, D.I.C., Goldblith, S.A., 1969. Dielectric loss factor of reconstituted ground beef: the effect of chemical composition. Journal of Food Technology 23 (944), 84–86.

Wang, Y., Wig, T., Tang, J., Hallberg, L., 2003. Dielectric properties of foods relevant to RF and microwave pasteurization and sterilization. Journal of Food Engineering 57, 257–268.

Zhang, Z., Su, T., , 1, Zhang, S., 2018. Shape effect on the temperature field during microwave heating process. Journal of Food Quality 2018. https://doi.org/10.1155/2018/9169875. Article ID 9169875, 24 pages.

Further reading

Nelson, S.O., Kraszewski, A., You, T., 1991. Solid and particulate material permittivity relationships. Journal of Microwave Power & Electromagnetic Energy 26 (1), 45–51.

CHAPTER 3

Microwave heating effects on foodborne and spoilage microorganisms

3.1 Introduction

The effectiveness of using microwave (MW) heating for pasteurization and sterilization of food and beverages was demonstrated and discussed over the decades in a numerous study (Laguerre and Hamoud-Agha, 2019). The objective of pasteurization process is to reduce vegetative pathogenic and spoilage organisms and to deactivate some enzymes in foods to enhance safety and extend product's shelf life at refrigerated conditions. Pasteurization temperatures and time combination can vary depending on the product, its pH, and the target organisms of concern. Typically, the food is heated up to 65–90°C for a time varying from 30 min to a few seconds. The objective of food sterilization process is to destroy bacterial spores in liquid, semisolid, or solid products in order to extend their shelf life at ambient storage conditions. The heat transfer in traditional processes takes place mainly by conduction from the surface to the center of the product, and the temperature of the process is measured in the "cold" spot with the lowest temperature. This can result in more severe heating conditions to reach the target temperature in the cold spot and consequently cause overcooking of the surface and a degradation of the quality of processed products. Therefore, optimizing thermal treatments (i.e., maximizing microbial inactivation while minimizing nutrient degradation) is a key issue and technical challenge in thermal processing.

Microwave heating offers an alternative to traditional pasteurization and sterilization processes for both solid foods such as prepacked ready-to-eat (RTE) meat and poultry products and liquid foods, such as milk and other dairy products, high and low acid juices, and other beverages. According to Mudgett (1989), higher product quality and extended shelf life at refrigerated and ambient temperatures can be achieved with microwave heating due to their volumetric nature and higher heating rates and consequently reduced processing times (PTs) as compared to conventional thermal processes. Microwave food processing systems can be more energy-efficient, and the system can be turned on and off instantly, so operation is easily controllable. Microwave heating permits superior retention of moisture and macronutrients (Ruello, 1987). Nutrient retention is enhanced as a result of rapid come-up time and shorter duration of heating (Datta and Hu, 1992). Microwave sterilization of prepackaged foods is several times faster than heat sterilization processing (Morris, 1991). Microwave systems can be easily cleaned, and there is no problem of product fouling on the surface of the equipment, thus resulting in reduced water use.

Numerous studies have documented the effective destruction of pathogenic and spoilage microorganisms and enzymes during microwave heating in food products (CFSAN/IFT, 2000). Microorganisms shown to be inactivated by microwaves include both spores, vegetative bacteria, molds, yeasts, and viruses including *Bacillus cereus, Clostridium perfringens, Clostridium sporogenes, Campylobacter jejuni, Escherichia coli, Listeria monocytogenes, Staphylococcus aureus, Salmonella, Enterococcus, Saccharomyces cerevisiae, and Aspergillus*. From the available literature data, however, it is challenging to accurately estimate accurately the effectiveness of microwave heating systems because different techniques and methodological approaches are employed and appropriate detailed descriptions of temperature monitoring methods and process uniformity are generally absent. Another reason for challenge in comparing the efficiency of microwave and conventional heating process is in the inherent physical differences of heating modes between the two technologies: direct absorption of microwave energy versus conduction and resulting variance in heating rates. Absorption of energy of the high frequency electromagnetic field by the materials with dielectric losses of such foods is known to be a complex multiphysical phenomenon which, in addition to electromagnetic and heating processes, may involve evaporation, mass transfer, and other phenomena. Due to this complexity of microwave interactions with food matrix, simply placing a sample food in a microwave cavity will rarely lead to success. As a result, vast majority of studies of microwave food heating for pasteurization and sterilization processes used systems that were not designed or optimized with the heating pattern dependable on a large number of critical factors such as food itself, packaging, location, and heating system.

Despite the numerous advantages of microwave decontamination, heating nonuniformity is a drawback that leads to incomplete inactivation of microorganisms. Additionally, the nonuniform heating causes a severe deterioration of food quality because local overheating can result in quality changes such as color or evaporation in locations where the temperature is the highest. Frequently, these phenomena can be detected at the corners and the edges of the product due to microwave reflection.

Another challenge concerning microwave heating is an information about cold spot in the treated product. During a typical thermal process, the cold point is well defined and located in the one point of the geometrical center of the product. During microwave pasteurization, one-point temperature monitoring within the product is not sufficient to ensure food safety.

Regularly, measuring the internal temperature during microwave treatment did not reflect the surface inactivation, where the temperature was lower.

Based on the above, temperature measurement and control is of crucial importance in microwave pasteurization and sterilization, especially with respect to temperature distribution, process uniformity, rate of temperature rise, final temperature reached, and knowledge of the fastest and slowest heating points inside and on the surface of the product. To locate the cold spot, the chemical marker method developed by Kim and Taub was successfully used in the development of microwave sterilization process (Kim and Taub, 1993). Combined with experimental investigations, modeling and numerical simulations are highly recommended to find and validate the product cold spot. This use of modeling methods will allow to accomplish more accurate approach and to develop a reliable microwave process. Additionally, in order to design and optimize a reliable microwave pasteurization and sterilization process, electromagnetic energy transfer models need to be coupled with heating model and microbial destruction kinetics model. The resulting multiphysics model has to describe and predict the spatial temperature distribution as well as the microbial inactivation at every point within the product sample.

Additionally, the increased utilization and implementation of microwave heating processing has been inhibited by insufficient knowledge on the kinetics of microbial destruction during microwave heating and approaches to evaluate it. Lack of reliable temperature control during microwave heating of heterogenous foods has fruitless efforts to obtain kinetic data. Hence, a fair comparison cannot be made with destruction kinetics data obtained during conventional heating. There are very few reliable data on the effect of microwave heating rate on microbial destruction kinetics. The uncertainty in much of the published literature creates difficulties in the formation of general conclusions about microwave heating and its application for design and validation of pasteurization or sterilization processes.

This chapter will discuss the differences in the evaluation of microbial destruction kinetics between conventional and microwave heating, reported microwave effects on different types of microorganisms including spores, vegetative bacteria, yeasts, molds, and viruses, microbial microwave thermal resistance, and effects of food composition. The possible existence of nonthermal (athermal) effects of microwaves is a subject of theories regarding interaction between microwaves and certain cellular constituents. Because this is a subject of extensive continuous debate, it won't be discussed in this chapter.

3.2 Kinetics of microbial inactivation under microwave heating

Critical components of the development of microwave preservation process are shown in Fig. 3.1. As discussed above, the performance objective of microwave preservation process (shelf life extension, pasteurization, or sterilization) is to achieve target microbial reduction. This means that any microwave preservation process cannot be developed without the sufficient knowledge of the kinetic parameters of microbial inactivation or enzymatic destruction and understanding attenuation or promoting effects of food matrix, pH, water activity, and chemical preservatives.

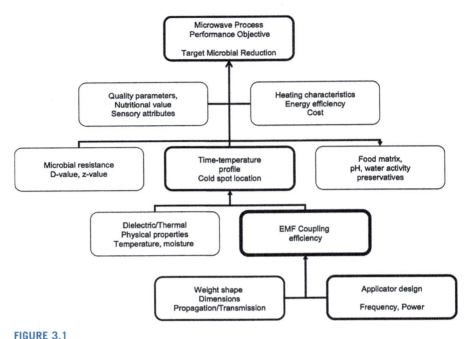

FIGURE 3.1

Components of the development of microwave food preservation process.

Another process objective such as higher product quality and superior retention of nutrients can be achieved as a result of rapid come-up time and shorter duration of process as compared to conventional thermal processing.

Kinetics of microbial inactivation and heat resistance during conventional thermal processing traditionally has been described by a first-order reaction kinetics and characterized by means of D-value or decimal reduction time (min) and Z-value or relative heat resistance, °C. These two values are fairly available for different groups of microorganisms, various heating modes (dry or wet) and are well documented and recommended for use in the design of microwave heating processes.

In the absence of any nonthermal effects, the inactivation of microorganisms and destruction of enzymes are generally modeled as nth order chemical reaction described as Eq. (3.1):

$$dC/dt = -kC^n \tag{3.1}$$

in which dC/dt is the time rate of change concentration C (CFU/mL)/s; k is the reaction rate constant (1/s); n is the order of reaction. For many foods, a first-order kinetics model adequately describes the microbial inactivation in the processes where high heating uniformity can be achieved.

As already mentioned, the thermal resistance of microorganisms at reference process temperature T_{ref} is traditionally characterized in the food processing by means of the D-value (at T_{ref}) and Z-values calculated using Eq. (3.2) (Stumbo, 1973):

$$D_{T_{ref}} = 2.303/k$$

$$z = \frac{T_2 - T_1}{\log D_1/D_2} \quad (3.2)$$

When the thermal resistance of a microorganism is known, it is possible to calculate the equivalent time or process lethality value or F-value that is necessary for thermal treatment at reference temperature by integration of the measured time-temperature history in the product using Eq. (3.3).

$$F = \int_0^t 10^{\frac{T(t)-T_R}{z}} dt \quad (3.3)$$

This approach has been traditionally used in the thermal process calculations and process design to achieve the required lethality in the product. Similar concept can be applied for determining kinetics parameters during microwave heating; however, nonisothermal microwave heating conditions are involved in this case. Microwave effective time at reference temperature has to be evaluated and used in the calculation of D-value according to Eq. (3.4). Then, resulting D-values can be computed as:

$$D = t_{eff}/\log(C_o/C) \quad (3.4)$$

where t_{eff} is an effective heating time (same as F in Eq. 3.3) with T_R as exit or reference temperature, obtained using either model predicted or experimentally determined time-temperature profiles; C_o and C are initial and final concentration of microbial cells (CFU/mL).

The use of this approach is rather rare in studies concerning microwave effects. There have been only a few studies describing kinetics during microwave heating and most results refer to the evaluation of lethality effects without considering thermal history of the product (Riva et al., 1993). Similar to thermal destruction, microwave destruction kinetics of food constituents such as quality attributes, enzymes, and vitamins are required in addition to microbial destruction for establishing and validating microwave process.

Table 3.1 shows the application of this approach when D-values of *E. coli* K1 2 bacteria have been estimated from the experimental microbial survival data in buffer at different temperatures 55–65°C based on the computed effective times from time-temperature data (experimental or perfect mixed flow model predictions) for microwave heating, hot water, and steam heating modes conditions (Le Bail et al., 2000). A comparison of data between the D-values in thermal batch heating, microwave, steam, and hot water continuous-flow heating shows that lowest D-values were observed with the continuous-flow microwave heating condition and ranged

Table 3.1 Experimental D-values of *Escherichia coli* K-12 bacteria in buffer during microwave heating and conventional heating in different batch and continuous modes.

Temperature (°C)	D-values (s)		
	55	60	65
Thermal batch method	173.0	18	1.99
Continuous-flow, hot water	44.70	26.80	2.00
Continuous-flow, steam	72.71	15.61	2.98
Continuous-flow, microwave	12.98	6.31	0.78
Continuous-flow, microwave + holding	19.89	8.33	1.98

from 13 s at 55°C to 0.78 s at 65°C. The experimental D-values in microwave system with hold time of 20, 8.3, and 2.0 s at 55, 60, and 65°C were also lower than those obtained for steam heating (73, 16, and 3.0 at 55, 60, and 65°C) and hot water heating. Further, D-values in continuous systems were considerably lower than those in batch heating systems that indicated the critical role of mixing in delivering required temperature.

In practice, in order to evaluate the microbial lethality at the coldest spot of the product, the kinetic inactivation data of organisms of concern are required. Much of the data reported cannot be considered to accurately represent the kinetics of microbial destruction solely during microwave heating since they also include the effects of the hot-holding or cooling period following microwave heating.

Laguerre et al. (2011) proposed new specific power destruction parameters, D_p- and Z_p-values, to describe microwave microbial inactivation. The decimal reduction time (D_p value) is the treatment time required to reduce microbial population by 90% at a constant applied specific power expressed in W/mL or in W/g. The Zp value is the change of specific power necessary to cause a tenfold change in the D_p values of microorganism under specified conditions. This concept has been successfully tested for the sterilization of infant food in a lab-scale microwave pilot unit and allowed to determine the optimal condition for microwave sterilization of infant food.

Because of the nonuniformity of microwave heating, microbial inactivation kinetics need to be coupled with nonlinear heat transfer model to calculate the temporospatial survival of bacteria (Hamoud-Agha et al., 2013). Mallikarjunan et al. (1996) developed a mathematical model that included mass heat transfer to microbial inactivation kinetics. The variation of the dielectric properties with respect to the temperature has been considered in the simulation process, and a good agreement with experimental data was obtained.

3.3 Effect of microwave energy on microorganisms

Since 1940s, when work on the destruction of microorganisms by microwave heating was reported by Fleming (1944), numerous studies documented the effectiveness of microwave heating on microbial inactivation and attracted a considerable interest. Since that time, a range of potential pathogens and spoilage microorganisms and enzymes has been inactivated by microwave heating of food products.

As discussed earlier, many of the reported studies of microwave pasteurization and sterilization processes have used systems not designed or optimized for the purpose due to the complexity of the microwave heating process, where heating patterns depend on a large number of critical factors. Nonuniform and uncontrolled MW coupling has been a key problem during many of these studies.

Table 3.2 provides some selected data from the published studies which had sufficient data on temperature control and inactivation kinetics parameters.

Different Z-values for *Enterobacter cloacae* and *Streptococcus faecalis* in batch conventional heating (4.9 and 5.8°C) and microwave heating (3.8 and 5.2°C) have been reported. The results were explained rather due to different heating kinetics and nonuniform local temperature distributions during microwave heating than existence of specific athermal effects.

Luechapattanaporn et al. (2004) validated radio frequency sterilization of mashed potatoes in trays. Thermal inactivation kinetic parameters were used to calculate a process lethality. The microbial challenge studies data showed that microbial destruction from the validated process agreed with the calculated F-values of sterilization process.

The majority of these data reported log kill achieved in the specific products, under particular heating conditions to demonstrate overall effectiveness of microwave heating against pathogenic bacteria such as *E. coli*, Listeria, and Staphylococcus. Only a few studies (Tajchakavit et al., 1998) evaluated D-values of *Saccharomyces cerevisiae* and *Lactobacillus plantarum* and showed higher microwave sensitivity *of L. plantarum* bacteria at 52.5 and 55°C with decreasing sensitivity at higher temperature of 57.5°C.

There do not appear to be any obvious reported "microwave-resistant" foodborne pathogens. As with conventional heating, bacterial organisms are more resistant to thermal inactivation by microwave heating than yeasts and molds, and bacterial spores are more resistant than vegetative cell and can be ranked as:

Microbial spores > vegetative cells > yeasts and molds

The efficacy of microwave heating on the viability of *Cryptosporidium parvum* oocysts and on the sporulation of *Cyclospora cayetanensis* oocysts for various periods of cooking time at 100% power at 700 W was determined by Ortega and Liao (2006). *Cryptosporidium parvum* oocysts could be completely inactivated as

Table 3.2 Inactivation effects of microwave and radio frequency heating on selected microorganisms in various foods.

Microorganism	Microwave conditions	Heating Medium Product	Inactivation or kinetic data	Source
Escherichia coli	2450 MHz 55°C, 30 s	Frozen shrimp	3-logs kill	Odani et al. (1995)
Escherichia coli K12	Microwave (MW) oven 2450 MHz, 600 W 50 and 60°C	0.9% NaCl solution	3-logs kill	Woo et al. (2000)
Staphylococcus aureus	55°C, 30 s	Frozen shrimp	3-logs kill	Odani et al. (1995)
Staphylococcus aureus	MW oven 2450 MHz, 800 W 61.4°C, 110 s	Steel disks	6-logs kill	Yeo et al. (1999)
Bacillus cereus spores	100°C, 90 s cells 100°C	Frozen shrimp	3-logs kill	Odani et al. (1995)
Clostridium sporogenes PA 3679	Radio frequency heating 27.12 MHz T = 121.1°C 25 min	Mashed potatoes	>4.1-logs kill	Luechapattanaporn et al. (2004)
Bacillus subtilis KM 107cells	MW oven 2450 MHz, 600 W 60–80°C	0.9% NaCl solution	2-logs kill	Woo et al. (2000)
Aspergillus nidulans spores	MW oven. 2450 MHz, 700 W 30 min, 85°C	Glass beads	3-logs kill	Diaz-Cinco and Martinelli (1991)
Listeria monocytogenes	MW oven, 2450 MHz, 600 W Isothermal 75, 80, and 85°C	Beef frankfurters	75°C—8-logs after 13.5 min 80°C—8-logs After 11–13 min 85°C—8-logs after 12 min	Huang (2005)

3.3 Effect of microwave energy on microorganisms

Listeria monocytogenes Scott A	MW oven, 2450 MHz 750 W	Nonfat milk	$D_{60} = 228$ s $D_{82.2} = 0.57$ s 4–5 logs at >71.1 for 15 s	Galuska and Vasavada
Saccharomyces cerevisiae	MW, 2450 MHz, 700 W Continuous flow 52.5, 55, and 57.5°C	Apple juice	$D_{52.5} = 4.8$ s $D_{55} = 2.1$ s $D_{57.5} = 1.1$ s $Z = 7$°C	Tajchakavit et al. (1998)
Lactobacillus plantarum	MW, 2450 MHz, 700 W Continuous flow 52.5, 55, and 57.5°C	Apple juice	$D_{52.5} = 14$ s $D_{55} = 3.8$ s $D_{57.5} = 0.79$ s $Z = 4.5$°C	Tajchakavit et al. (1998)
	MW oven, 650 W 50°C, 30 min	Culture broth	4-logs reduction	

little as 20 s of cooking time, whereas Cyclospora sporulation was observed up to 45 s. *Cryptosporidium parvum* could be inactivated only when temperatures of 80°C or higher were reached in the microwave ovens.

3.4 Effects of food properties on microbial inactivation under microwave heating

Intrinsic food properties of various products are those that belong within the product itself. These parameters are pH, water activity (a_w), redox potential, nutrient content, antimicrobial constituent, and structure. They can be altered when undertaking new product development or when aiming a product at a different market (e.g., low sugar, low salt, low calorie, or high energy). Most of these intrinsic factors will affect the ability of any process to inactivate microorganisms. The composition of foods such as its water content, salts, sugar, proteins, fat content, and product structure is critical in understanding microbial inactivation in microwave processing.

The association between product pH and thermal process has long been exploited for shelf-stable canned foods. This is based on a requirement to prevent growth and toxin formation by *Clostridium botulinum spores*, where growth can be prevented in product with pH 4.6 or lower. In such high acid and acidified foods, a pasteurization designed to kill vegetative organisms and control of aciduric spore-forming bacteria can be used. If the product pH is higher than 4.6, *C. botulinum* may grow and a lethal "botulinum cook" must be employed. Reducing the pH decreases the thermal process required to destroy *C. botulinum* in canned foods. For commercial canning of low acid (pH > 4.6, a_w > 0.85) shelf-stable foods, *C. botulinum* type A and B (proteolytic) spores are the target bacteria; the corresponding Z-value is 10°C (Stumbo, 1973), T_{ref} is taken as 121.1°C, and canning thermal processes are designed to achieve Fo = 3 min, corresponding to a minimum of 12-log reduction in *C. botulinum* type A and B spores.

Microbial validation of microwave sterilization process was conducted using *C. sporogenes* PA 3679 spores (Guan et ai., 2002, 2003). Thermal resistance (D- and Z-values) of PA 3679 cultures were first determined in M/15 Sorensen's phosphate buffer (pH 7.0) to calibrate surrogates. Only spore crops with a Z-value close to 10°C (the same as *C. botulinum type A and B* spores) and D121.1°C (D-value at 121.1°C) values over 0.6 min (about 2.5 times longer than that of *C. botulinum spores type A and* B) were used for microbial validation tests. Then tests were conducted to determine thermal resistance of the selected spore crop in different food components (for example, sauce and salmon fillets) to be processed in the same packages. The food component in which the calibrated PA 3679 spore crop had the highest D121.1°C value and also had the slowest heating at cold spots was inoculated for microbial validation tests.

In acid foods, the concern is in reducing vegetative pathogens of concern through effective process controls. There is considerable literature indicating that lowering the product pH increases the lethal effects of temperature that also relates to microwave heating. For example, the targeted bacteria may be a vegetative pathogen such as *L. monocytogenes*, and T_{ref} may be a selected temperature between 65 and 90°C. Six-log reductions are often used in process calculation.

According to Heddleson et al. (1996), the heat resistance *of Salmonella, S. aureus,* and *L. monocytogenes* is greater at their optimum growth pH. When the product pH was in the range of 6.8–7, there was no significant difference in the survival of Salmonella species.

Reducing water activity (a_w) is a well-known preservation technique for preventing microbial growth. Due to the importance of this factor, the term "low-moisture foods" has been introduced to describe products with a_w lower than 0.85, which refers to a threshold below which bacterial pathogens would not be able to grow in those products. The reduction of a_w won't kill microorganisms, but it will assist to control their growth. Pathogens generally will not be able to multiply or produce toxins in foods with a_w <0.85. However, when low-moisture foods or ingredients are used in high-moisture finished products, growth conditions for pathogens are reinstated. The protective effects of low a_w have been shown in thermally processed foods and observed in other types of processes including microwave heating; however, the effect may differ when different solutes (e.g., sodium chloride, glycerol, fructose, and sorbitol) are used (Setikaite et al., 2009) and impact dielectric properties. This effect shall be considered during the validation of microwave-based processes.

Heddleson et al. (1996) examined the chemical composition of five foods to determine the important factors in achieving uniform temperature within them when heated in a 700 W microwave oven. The chemical composition of foods was related to temperature and destruction of microwave-heated *Salmonella, L. monocytogenes,* and *S. aureus*. Thermotolerance was found to be strongly influenced by such food components as lipids, salts, and proteins. Chipley (1980) noted that *Staphylococci* were recovered after microwave heating over a range of 35–92°C indicative of inadequate thermal treatment. The results indicated that the level of destruction of *Salmonella, L. monocytogenes,* and *S. aureus* within foods varied significantly. The influence of food composition on bacterial inactivation differs, and it is more pronounced for *Salmonella* and to a lesser degree for *L. monocytogenes*. For example, foods with higher sodium content exhibit surface heating and nonuniform temperature distribution which favors the survival of bacteria. In general, it was found that bacterial inactivation by microwave heating is decreasing with increasing salt content. Dissolvent salts decrease the penetration depth of incident microwave energy, and this results in inhomogeneities in temperature distribution. In cases of high salt content, the surface temperature is much higher than in the rest of the sample. When the moisture content is high, products will heat more efficiently due to the larger dielectric loss factor. But as moisture content decreases, the penetration depth of the microwave energy increases and products can still heat well if they also have low heat capacity

(Heddleson et al., 1994). Moisture and salt are principal determinants of a food's dielectric behavior. Conclusively, by contrasting the temperature profiles and the level of bacterial destruction, it is obvious that the mixed average final temperature cannot be used as an indicator for microbial destruction.

In a different experiment, Heddleson et al. (1991) examined the protective effects that food constituents may exert on *Salmonella* species when heated by microwave energy. The operating frequency of the oven they used was at 2.45 GHz and the power was 700 W. The samples were heated for 47 s, and the results showed that the sodium chloride provides most of the protection with 37.6% of the *Salmonella* population surviving when on the contrary in the absence of salt only 0.22% of the population survived. The presence of salt drastically influences the uniformity of temperature within the sample and results in surface heating. Heddleson et al. (1994) concluded that there are several factors that affect the survival of *Salmonella* in different microwave ovens. When the average final temperature was less than 57°C or less, there was significant less bacterial destruction ($\sim$85%) compared to 60°C (95%). When milk was heated for a constant time of 45 s at high, medium, and low power settings, samples heated at the highest settings had significant more destruction of *Salmonella*. Microwave heating to a common mean final temperature resulted in no significant difference between destruction in containers of different shape with different volumes. Heating to an end-point temperatures (EPTs) is a better means to achieve consistent microbial destruction.

3.5 Effects on microwave heating parameters on microbial inactivation

The results of the studies also suggest that microwave heating is a successful alternative to conventional continuous pasteurization of liquid foods, such as juices, milk, juices, and other beverages and liquid products with high viscosity and particulates.

3.5.1 Microwave continuous flow systems
3.5.1.1 Juices
USDA-ARS ERRC scientists have conducted several studies into radio frequency electric field (RFEF) technology, subjecting liquid foods to its high electric fields, and found it to be both efficient for pathogen inactivation and cost-effective. In one study, scientists applied RFEF for 3 s at 60C to an apple juice sample inoculated with *E. coli*. The electrical cost of RFEF processing was about one cent per decaliter (10.5 quarts), and the procedure was more effective than conventional heating under the same conditions.

Genry and Roberts (2005, USDA-ARS, Pacific Research Center) reported about the design and evaluation of a continuous flow microwave pasteurization system (input power 900–2000 W) for the pasteurization of apple cider. Volumetric flow

3.5 Effects on microwave heating parameters on microbial inactivation

rates and absorbed power were criteria for evaluation. Process lethality was verified based on the inactivation E. coli 25,922 in apple cider and pasteurization process resulted in 5-log$_{10}$ reduction. The effect of product viscosities, load size (0.5 and 1.38 L), inlet temperatures, and flow characteristics were evaluated. No significant difference in absorbed power with respect to levels and position of coils within the microwave cavity was found. It was reported that any volume greater than 1 L in this microwave oven will heat at the rate determined by the input power of the oven. The inlet temperature had a significant effect on the flow rates of the fluids because of the effect on product viscosity. However, the initial heating rate was greater for the cider entering at 3°C followed by the cider entering at 21°C, and the cider entering at 40°C had the slowest heating rate. This was explained by the fact that dielectric properties for food without salt are greatest at temperatures just above the melting point. It was concluded that the application of microwave energy for the pasteurization of apple cider was a feasible thermal process, especially with increased power. The major factor in the design of microwave system is to ensure that the fluid is obtaining uniform thermal energy.

The effectiveness of microbial destruction was measured under continuous-flow microwave heating conditions by inoculating apple juice from Motts (pH 3.5, Brix 16.5) with *E. coli K12* to a concentration of 10^6 CFU/mL. It was recommended that for practical purposes, *E. coli K12* surrogate of *E. coli O157:H7* could be considered as target microorganism for which a 5-log population decrease must be achieved. The results of the tests are shown in Table 3.3. It was found that microwave heating at temperature of 57–71°C was effective in killing *E. coli K12* in apple juice. 99.55% of bacterial inactivation was achieved at final temperature of 57.2°C, and 100% of inactivation was achieved at final temperature of 63.4°C. Treatment time of 28.5 s of microwave heating to a temperature of 63.4°C resulted in a 6-log reduction of *E. coli* bacteria in apple juice.

Kozempel et al. (2000) indicated the possibility of a synergistic interaction of the microwave energy and some component in the apple juice that injuries or kills

Table 3.3 Inactivation of *Escherichia coli* K12 by microwaves in apple juice at initial temperature $T_o = 22°C$.

Mass flow rate (g/s)	Exit temperature (°C)	Residence time (S)	Survival ratio (N/No)	Inactivation (%)	Log$_{10}$ (N/No)
11.5	57.2	25	0.453	99.55	0.343
11	60	27	0.112	99.888	0.949
10.5	62.8	28	0.000205	99.9998	3.689
10	63.4	28.5	5.298E−07	100	6.276
9	65	29.5	0	100	
7	71	40.6	0	100	

microorganisms. According to the author, the suspected component can be a malic acid. However, uneven heating, irregular heating patterns in foods of varying heterogeneity, and much shorter heating times encountered when using microwave energy can result in the survival of bacteria, even when the overall temperature in the product can be adequate to kill microorganism. In such cases, the adequate holding time need to be provided at the end of microwave heating cycle to eliminate hot and cold spots and to allow for the equilibration of local temperatures within the food product.

3.5.1.2 Milk and protein beverages

Lopez-Fandino et al. (1996) compared thermal treatment of milk during continuous microwave (CW) and conventional heating. It was found that microwave treatment of milk is a mild and efficient way of providing a product with satisfactory microbial and sensory qualities without causing extensive heat damage.

Knutson et al. (1988) simulated the heat treatment given to milk at high-temperature, short-time (HTST) process and low-temperature, long-time (LTLT) pasteurization (71.1°C for 15 s and 62.8°C for 30 min, respectively) in a microwave oven. They investigated the heat resistance of *Coxiella burnettii, S. typhimurium, E. coli*, and *P. fluorescens*. Uninoculated steamed milk (76 mN) was heated, and the time and temperature parameters were described using linear regression. According to the equation of the line, a heating time of approximately 60 s corresponded to achieve the temperature of 72.8°C. Again, tested microorganisms were recovered in this condition. In a similar way, samples of 125 mN of inoculated milk were heated, and according to the equation of the line, 80 s of heating corresponded to 75.2°C. HTST pasteurization temperature of 75°C was reached in 76 s. However, when milk was heated for time periods in excess of that shown as necessary to reach 71.7°C, inoculated organisms were recovered indicative of a nonuniform distribution of heat through the food. LTLT pasteurization is defined as holding milk to 62.8°C for 30 min. Milk was inoculated with *S. faecalis*, and when heated to 62.8°C, the number of survivors decreased by 3 orders of magnitude (Knutson et al., 1988). The results showed that one of the 5 trials did not reach this goal probably because of uneven temperature distribution in the milk. Nikdel et al. (1993) reported about complete destruction of *L. plantarum* bacteria and enzyme of pectin methylesterase in orange juice without changing the taste in the lab-scale CW pasteurizer. Tajchakavit et al. (1998) explored the application of microwave energy for the destruction of spoilage yeasts of *S. cerevisiae* and *L. plantarum* lactic bacteria and suggested enhanced thermal effects associated with microwaves.

Clare et al. (2005) designed and implemented CW thermal processing of skim fluid milks (white and chocolate) to compare sensory, microbiological, and biochemical parameters with indirect ultra-high temperature (UHT)-treated milks. No-Bac Unitherm model Cherry Burtell Corporation unit was used for UHT treatment, whereas a 60 kW continuous flow MW heating unit from IMS at 915 MHz was used for MW heating. Milk was aseptically filled into 250 mL brick-style aseptic cartons. Specifically, bacterial counts, plasmin activity, DTNB reactivity protein sulfhydryl groups, sulfhydryl oxidize activity, viscosity, color parameters, and descriptive sensory tests wre performed. All test products were aseptically packaged and stored at ambient conditions for 12 months. The results of the study demonstrated that MW processing holds much promise for the future due to better sensory characteristics.

Bookwalter et al. (1982) from USDA-ARS (Peoria, IL) reported the results of microwave processing of corn-soy–milk blends to destroy Salmonellae and evaluate effects on product quality. The product was heated from 21°C to 56.7 through 82.2°C in 3.9–10 min using a 60 kW (2450 MHz) CW tunnel. Palletized bags were cooled to 43°C after 9.7 h with moving air. The initial MPN numbers of 4×10^2 CFU/g for *Salmonella senftenberg* were reduced 10^2–10^5 fold after processing at 56.7–82.2°C, respectively. The samples were tested for physicochemical parameters, color changes, vitamin A and B, and available lysine. No statistically significant changes in quality were found at process temperatures 67°C and below.

Adequacy of pasteurization of milk using MW at 2450 MHz and 1.5 kW generator was measured by phosphatase tests and bacterial plate counts. Milk was pasteurized at 3 flow rates of 200, 300, and 400 mL/min. Treatment was conducted at 72°C for 15 s. In all instances, phosphatase test was negative. A triangle taste panel found no significant difference between microwave-pasteurized milk and the control. Results demonstrated the applicability of microwave energy as a heat source for continuous pasteurization of milk (Jaynes, 1975).

3.5.2 Viscous products and products with particles
3.5.2.1 Sweet potatoes

Sweet potatoes puree (SPP) was aseptically processed using a continuous flow microwave unit (Industrial Microwave Systems [IMS], Morrisville, NC, USA) to obtain a shelf-stable product. The processing conditions were established in the lab-scale unit consisted of a 5 kW MW generator, operating at 915 MHz a waveguide of rectangular cross section with directional coupler and specially designed applicator. A tube of 1.5-inch nominal diameter made of Teflon was placed at the center of the applicator. SPP was pumped at a rate of 0.5 L/min. The process was validated in a 60 kW continuous flow microwave unit from SPP and was aseptically processed using a continuous flow microwave unit (Industrial Microwave Systems [IMS], Morrisville, NC, USA). The flow rate was 4.0 L/min. The product was heated to 135 and 145°C, held for 30 s and rapidly cooled in a tubular heat exchanger and then aseptically packaged in aluminum-polyethylene laminated bags using a bag-in box unit. The puree bags were stored at ambient temperatures. The pilot-scale tests produced a shelf-stable product with no detectable microbial count during a 90 days storage period. The results of this study were reported by Coronel et al. (2005).

To establish the feasibility of microwave heating for the sterilization of sweet potato purees in 2007, this microwave system validated by microbial inactivation studies using bioindicators spores of *Geobacillus stearothermophilus* ATCC 7953 and *Bacillus subtilis* ATC 35021. Sweet potato purees seeded with the bioindicators in flexible food-grade pouches were subjected to 3 levels of processing based on the fastest particles: undertargeted process $F_o = 0.65$ min, target process $F_o = 2.8$ min, and overtargeted process F_o around 10.10 min. The log reduction results for B. subtillis were equivalent to the predesigned degrees of sterilization (Brinley et al., 2007).

3.5.2.2 Salsa
The feasibility of aseptic processing of low acid salsa product was also showed using 5-kW continuous flow pilot-scale microwave system. The product was heated up to 130°C and showed narrow temperature distribution between the center and the wall of the tube. Dielectric properties of the product were also measured, and correlations with increase of temperature were found (Kumar et al., 2007).

The possibility of sterilizing heat-sensitive products at relatively low temperatures for a short time is very intriguing. If microwaves can sterilize dry products (like spores) more efficiently at low temperatures and long exposure times than convectional heating, then it can be assumed that this is due to the nonthermal effect of microwaves. Jeng et al. (1987) inactivated spores by exposing them to microwaves in the system of 4 kW at 2.45 GHz. The total exposure times to reduce 10^5 spores were 75 and 48 min at 130 and 137°C, respectively. Jeng et al. (1987) discovered that microwave heating at temperatures lower than 117°C was not effective, and this result suggests that the sporicidal activities of the microwaves are due to the thermal effect.

3.5.2.3 In-package sterilization and pasteurization
Microwave heating of prepared meals in sealed retail containers can be used to extend shelf life. The concept of prepared RTE meals using microwave heating to extend shelf life has been considered for many years, and pilot studies have shown the potential of this method (Burfoot 1990; Burfoot et al., 1996). However, there has been relatively little adoption of the technology by the food industry, partly due to the technical difficulties associated with the process.

3.5.2.4 Macaroni and cheese product in a single tray
A 915 MHz microwave-circulated water combination heating technology was validated for a macaroni and cheese product in a single tray using inoculated pack studies and spores of C. sporogenes (PA 3679) (Guan et al., 2003). Water immersion approach was used to improve the nonuniformity of MW heating. The pressurized microwave heating vessel allowed a treatment of a tray under an over pressure. The product in the vessel was preheated first to 75°C with circulated hot water at 100°C. The combined heating started when microwave power was turned on and the circulated water was set at 120°C. The holding stage was maintained by the circulated water while microwave was turned off. After the desired holding period, the tray was cooled for 2 min using water at 80°C under pressure and then by 20°C tap water at ambient temperature. The temperature of the product was controlled by fiber optic temperature sensors. To design the target process, the thermal resistance parameters of PA spores were determined and used. The results of the

study suggested that microbial destruction matched with designed degrees of F_o value and the technology has a potential in sterilizing packaged foods.

3.5.2.5 Bottled pickled asparagus

Lau and Tang (2002) used 915 MHz at 5 kW nominal power of microwave system (Microdry Model IV-5 Industrial Microwave generator) for heating pickled asparagus in 1.8 kg glass bottles. To achieve target temperature of 88°C, 15 min of heating was required at 1 kW output power and 9 min for 2 kW compared with 30 min of heating in conventional hot water. The pasteurization regime used by the industry is 74°C for 15 min followed by cooling. It is difficult to achieve rapid heating in the glass bottles using conventional thermal treatments such as hot water or steam. Peroxidase (POD) and texture testing was conducted. Heating uniformity was measured using fiber optical temperature sensors. The temperature at the bottom surface was higher than at other locations in the glass bottle especially for 2 kW MW power level. The POD activity was all negative for the conventional thermal treatments and for 1 kW MW treatment. For 2 kW MW treatment, only 70% of POD tests were negative. The nonuniformity of heating contributed to the residual POD activity. It was concluded if uniform MW heating of bottled asparagus could be achieved, pasteurization of pickled asparagus using MW energy can reduce the heating period by half compared with conventional heating.

3.5.2.6 Beef frankfurters

A microwave heating system equipped with a proportional—integral differential control device was developed for in-package pasteurization of RTE meats. The study demonstrated the feasibility of such a concept as a postlethality alternative to eliminate *L. monocytogenes*. However, edge heating effect was observed by shrinking of the plastic packaging material (Huang, 2005).

Clearly, the adequate design of a microwave processing system can dramatically influence the critical parameters of shelf life extension, pasteurization or sterilization processes such as the location and temperature of the coldest point to heat in the product. Before accepting microwave heating as a reliable method for pasteurization and sterilization of food, it is critical to ensure uniform heating during treatment. During a typical thermal process, the cold point is well defined and located in the geometrical center of the product. During microwave pasteurization or sterilization, one-point temperature monitoring within the product is not sufficient to ensure food safety. For example, Schnepf and Barbeau (1989) tested the inactivation of inoculated Salmonella in poultry and showed that measuring the internal temperature during microwave treatment did not reflect the surface inactivation, where the temperature was lower.

In this regard, the study of Huang (2005) is of special interest because computer-controlled microwave heating system was developed. The infrared sensor to monitor product surface temperature was utilized, and feedback control mechanism was used to control the power to the microwave heating oven. Results indicated that the simple on-off control mechanism was able to maintain the surface temperature of frankfurters near the respective set-point used in the study. This pasteurization process was able to achieve a 7-log reduction of Salmonella in inoculated beef frankfurters using a 600-W MW oven within 12–15 min at 75 and 80°C, respectively.

Welt and Tong (1993), Welt et al. (1994), Tong (1996) introduced an apparatus to evaluate the possible athermal effects of microwaves on biological and chemical systems and examined the limits and requirements to comparative studies between kinetics under microwave and conventional heating. They compared the inactivation of *C. sporogenes* and of thiamin under equivalent time-temperature treatments by perfectly stirred batch treatment by conventional and microwave heating and did not observe any athermal effect.

Lin and Sawyer (1988) examined the effects of wrapping food in polyvinylidene chloride (PVDC) film. The microwave power output of the heating system was of 713 and 356 W at 2450 MHz. They examined the survival of *S. aureus* and *E. coli* inoculated in ground beef assembled into a loaf, hence distributing the bacteria equally throughout the product volume. The effects of wrapping of the beef loaf in PVDC film, load size, microwave output power, and PT were chosen as experimental variables for this study. The authors introduced the concept of exposed microwave dose (EMD), defined as Watts x minutes processed/g of food, or EMD, to correlate wattage output, time of microwave processing, and load size. Immediately after treatment, the postprocessing temperature rise of the beef was measured. The results proved that the percent survival of three types of bacteria were lower after that had been wrapped in PVDC film than those that had not been wrapped. *S. aureus* survived better than *E. coli*. The effect of the PVDC wrap was less pronounced in the case of bigger load, but neither of the bacteria survived in the PVDC wrap after the longest exposure. When the bacteria were inoculated only onto the surface of food products and exposed to sufficient energy to be killed, then the PVDC film wrapping had no significant statistical effect. The use of 50% power (356 W) compared to 100% power for beef loaves did not affect the survival of *S. aureus* and *E. coli* when subjected to the same EMD. It was concluded that the number of surviving bacteria had an inverse relationship to PT and EMD but was directly related to EPT of the center of the beef loaf and microwave processing temperature rise. In practice, there is a challenge to accurately locate and measure the point of lowest temperature in food products processed by microwave energy. The relationship between parameters of EMD and EPT can be used to predict bacterial survival in microwaved beef loaf. It is important to note that PVDC film wrapping acts as a heat sink. It saves energy and improves the microbial quality of food, which can be improved even further if a standard EMD is calculated for microbial safety depending on the varying electrical properties and configurations. The effect of food shape on microbial survival and thermal response is still unclear, but it is possible that a standard shape of microwaveable food products can provide optimum control on bacteria destruction.

3.6 Combined action of microwaves with other chemical or physical factors

Microwave heating can also be beneficially combined with other energy sources and chemical preservatives. Koutchma (1998) and Koutchma and Ramaswamy (2000) has shown that combining microwaves with hydrogen peroxide in low concentrations can enhance the destruction of microorganisms. The synergistic effect was found through the calculation of the interaction coefficient as a ratio of the tested combined effect of two agents to theoretical additive inactivation effect. It was reported that the interaction coefficient had a maximum value at 0.075% of H_2O_2 and at T of 55/60°C in simultaneous application of hydrogen peroxide and microwave heating. It was concluded that by combining physical (microwave heating) and chemical agents (hydrogen peroxide), or with other hurdles, it is possible to reduce the severity of thermal treatment, while the overall impact on microorganisms still can remain high. The parameters of the agents (temperature of microwave heating and concentration of hydrogen peroxide) can be kept lower than under their individual use. The lower intensities or doses of utilized agents should result in minimizing the loss of product quality allowing the less extreme use of any single treatment.

Maktabi et al. (2011) reported the synergistic effect of microwaves with laser, and UV radiation *on E. coli* and other spoilage and pathogenic bacteria. It was found that the overall reduction in viable counts was significantly higher than the sum of the reduction values for the individual treatments. The order of the treatment processes had also a significant influence on microbial destruction. A successive process by laser, microwave, and then UV was the most effective.

Shin and Pyun (1997) investigated the effects of pulsed microwave (PW) radiation on cell suspensions of *L. plantarum*. The use of PW was proposed for pasteurizing foods with use of reduced heat, even though its biological effects were still not fully understood. When the pulse with the shortest time to be generated was used, it resulted in a very high peak power. In the case of application of short pulses with low repetition rate—the average microwave power can be relatively low, although the peak microwave power may be in mega Watts or giga Watt ranges. Pulse-modulated radiation may penetrate more deeply than continuous wave irradiation with the same frequency (Lin and Sawyer, 1988). The authors used a generator that was adjusted to deliver repetitive rectangular pulse trains at 500 pulses/second. The duty ratio was 1/10, equivalent to a pulse width of 0.2 ms and a period of time base of 2 ms. Experimental results showed that there was a 4-log cycle reduction of viable cells when exposed to a sublethal temperature of 50°C by CW irradiation, and at the same temperature, PW irradiation yielded lower viable bacterial counts by 2 or 3 orders of magnitude. Such findings indicate that PW is more effective against microorganisms. An indication of the heat damage of cells was the leakage of cell components into the heating medium. Under the influence of sublethal stress, cells experience a loss in membrane integrity, so amino acids, peptides, membrane lipids, and ions can be found in the extracellular environment as a response to injury. The injury of the *L. plantarum* cell membrane was more severe when PW was used, so probably the nonthermal effects of microwave irradiation could result in shifts across membranes and reorientation of long-chain molecules. The energy from microwave irradiation probably disrupts the membrane and/or subcellular structure, thus liberating lipids. Also, the lipid components of the cell membrane may be overheated during microwave irradiation and cause greater injury. The absorption of high-power PW radiation probably produces thermoelastic waves in biological tissues. Such pressure and displacement may cause physical damage to cell membranes and cytoplasm (Lin and Sawyer, 1988).

Fung and Cunningham (1980) reported that microwave heating in combination with conventional heating results in more uniform heating of food and better inactivation of bacteria. Datta et al. (2005) found that MW heating in combination with infrared heating or hot air jets decreases the nonuniformity of temperature distribution.

3.7 Conclusions

After decades of exploration of microwave heating for food applications, it has been shown its effectiveness against foodborne pathogens and spoilage organisms in low acid and acid foods and successful application in commercial food preservation processes. Food processors are creating a new generation of microwavable items due to consumers' increasing time pressure, nutritional awareness, and desire for foods that taste and smell like they were cooked in a conventional oven. In many cases, the products' success hinges on a combination of product reformulation, package design, and less processing. This new generation of microwave products includes meals, snacks, and everything in between, from fresh chicken and fish dishes, precooked entrees, and side dishes to grilled cheese sandwiches, biscuits, and pizza. The uniqueness of microwave sterilization and pasteurization technology is that it can be applied to both solid and liquid foods as well as complete meals sealed in multicompartment trays.

However, microwave process parameters and nonuniformity can negatively impact its effectiveness and process is still relatively poorly controlled because of complex interactions between foods and microwaves. A special attention should be given to product and packaging design and development. Furthermore, the heating heterogeneity is the major drawback of this technology. Several combined methods have been proposed to improve the process efficiency and heating homogeneity through coupling microwaves with other heating methods. These approaches can improve process control, microbiological safety, and quality of various products.

In 2011, the United States Food and Drug Administration (US FDA) approved Microwave-Assisted Thermal Sterilization Process (MATS) using 915 MHz. As reported, the technology immerses packaged food in pressurized hot water while simultaneously heating it with microwaves at a frequency of 915 MHz. This combination eliminates food pathogens and spoilage microorganisms in just 5—8 min and produces safe foods with much higher quality than conventionally processed RTEs. Wornick Foods, a manufacturer of convenience foods and customized meal solutions, was awarded a grant to establish MATS Research and Development Center at its facility in Ohio where consumer products companies can test the MATS process.

915 Labs has the worldwide license to manufacture and market microwave assisted sterilization and pasteurization technologies. According to 915 Labs, any food or beverage that will benefit from a lower processing temperature and a reduced PT is ideal for MATS processing. Heat-sensitive foods such as eggs, dairy ingredients, seafood, and pastas have all been successfully processed with MATS. The reported research demonstrated that compared to conventional thermal retorting, the shortened exposure to heat during microwave processing retains more nutrients, such as Omega 3s, B vitamins, vitamin C, and folate.

The company currently manufactures a system that perform both sterilization (MATS) and pasteurization (MAPS) treatments. The pilot-scale system is developed for product and process development. The smallest commercial production system processes 30 packages per minute. Larger capacity production systems with a throughput between 50 and 225 packages per minute are available by design. Actual capacities vary depending on the nature of the product, size of package, and the desired shelf life to be achieved. In order to realize the potential benefits of microwave processes and assist food companies to commercialize microwave technology and integrate it in the production process, 915 Labs offers product development, validation services, and packaging solutions.

Microwave continuous flow sterilization of homogeneous and particulate-containing high and low acid viscous products has been developed using frequency of 915 MHz. The US FDA acceptance has been also granted and cleared ways for commercial applications for pumpable low acid foods.

The future of microwave processing of foods appears to be the strongest for special applications, and it will probably be of limited usefulness as a general method of producing process heat. Microwave-based pasteurization and sterilization may be the technology that will enable ecommerce providers to deliver high-quality food directly to consumers through their traditional distribution channels.

References

Bookwalter, G., Shulka, T., Kwolek, W., 1982. Microwave processing to destroy Salmonella in corn-soy-milk blends and effect on product quality. Journal of Food Science 47, 1683—1686.

Brinley, T., Dock, C., Truong, V., Coronel, P., Simunovic, J., Sandeep, K., Cartwright, G., Swartzel, K., Jaykus, L., 2007. Feasibility of utilizing bioindicators for testing microbial inactivation in sweet potato purees processed with a continuous–flow microwave system. Journal of Food Science 72 (5), E 235–E242.

Burfoot, D., 1990. Microwave processing of prepared meals. In: Proceedings of Leatherhead Food Research Association Symposium "Microwaveable Product Technology" Symposium Series, vol 44, pp. 77–95.

Burfoot, D., Railton, C., Foster, A., Reavell, S., 1996. Modeling the pasteurization of prepared meals with microwaves at 896 MHz. Journal of Food Engineering 3, 117–1333.

Chipley, J.R., 1980. Effects of microwave irradiation on microorganisms. Advances in Applied Microbiology 26, 129–145.

Clare, D., Bang, W., Cartwright, M., Drake, M., Coronel, P., Simunovic, 2005. Comparison of sensory, microbiological, and biochemical parameters of microwave versus indirect UHT fluid skim milk during storage. Journal of Dairy Science 88, 4172–4182.

Coronel, P., Truong, V.D., Simunovic, J., Sandeep, K., Cartwright, G., 2005. Aseptic processing of sweet potato purees using a continuous flow microwave system. Journal of Food Science 70, E531–E536.

Datta, A.K., Hu, W., 1992. Optimization of quality in microwave heating. Food Technology 46 (12), 53–56.

Datta, A.K., Geedipalli, S.S.R., Almeida, M.F., 2005. Microwave combination heating. Food Technology 59, 36–40.

Diaz-Cinco, M., Martinelli, S., 1991. The use of microwaves in sterilization. Dairy Food Environment Sanitation 11, 722.

Fleming, H., 1944. Effect of high frequency on microorganisms. Electrical Engineering 63 (18).

Fung, D.Y.C., Cunningham, F.E., 1980. Effect of microwaves on microorganisms in foods. Journal of Food Protection 43, 641–650.

Gentry, T.S., Roberts, J.S., 2005. Design and evaluation of a continuous flow microwave pasteurization system for apple cider. Lebensmittel-Wissenschaft & Technologie 38, 227–238.

Guan, D., Plotka, V.C.F., Clark, S., Tang, J.M., 2002. Sensory evaluation of microwave treated macaroni and cheese. Journal of Food Processing and Preservation 26 (5), 307–322.

Guan, D., Gray, P., Kang, D.H., Tang, J., Shafer, B., Ito, K., Younce, F., Yang, T.C.S., 2003. Microbiological validation of microwave-circulated water combination heating technology by inoculated pack studies. Journal of Food Science 68 (4), 1428–1432.

Hamoud-Agha, M.M., Curet, S., Simonin, H., et al., 2013. Microwave inactivation of *Escherichia coli* K12 CIP 54.117 in a gel medium: experimental and numerical study. Journal of Food Engineering 116, 315–323.

Heddleson, R.A., Doores, S., Anantheswaran, R.C., Kuhn, G.D., Mast, M., 1991. Survival of *Salmonella* species heated by microwave energy in a liquid menstruum containing food components. Journal of Food Protection 54, 637–642, 67.

Heddleson, R.A., Doores, S., Anantheswaran, R.C., 1994. Parameters affecting destruction of *Salmonella* spp. by microwave heating. Journal of Food Science 59, 447–451.

Heddleson, R.A., Doores, S., Anantheswaran, R.C., Kuhn, G.D., 1996. Viability loss of Salmonella species, *Staphylococcus aureus*, and *Listeria monocytogenes* in complex foods heated by microwave energy. Journal of Food Protection 59, 813–818, 12.

Huang, L., 2005. Computer-controlled microwave heating to in-package pasteurize beef frankfurters for elimination of *Listeria monocytogenes*. Journal of Food Process Engineering 28, 453–457.

IFT, 2000. Kinetics of Microbial Inactivation for Alternative Food Processing Technologies. http://vm.cfsan.fda.gov/~comm/ift-pref.html.

Jaynes, H.O., 1975. Microwave pasteurization of milk. Journal of Milk and Food Technology 38 (7), 386–387.

Jeng, K.H.D., Kaczmareck, K.A., Woodworth, A.G., Balasky, G., 1987. Mechanism of microwave sterilization in the dry state. Applied and Environmental Microbiology 53, 2133–2137.

Kim, H.J., Taub, I.A., 1993. Intrinsic chemical markers for aseptic processing of particulate foods. Food Technology 47, 91–97.

Knutson, K., Marth, E.H., Wagner, M., 1988. Use of microwave ovens to pasteurize milk. Journal of Food Protection 51, 715–719.

Koutchma, T., Ramaswamy, H.S., 2000. Combined effects of microwave heating and hydrogen peroxide on the destruction of *Escherichia coli*. Lebensmittel-Wissenschaft und-Technologie 33, 21–29.

Koutchma, T., 1998. Synergistic effect of microwave heating and hydrogen peroxide on inactivation of microorganisms. Journal of Microwave Power & Electromagnetic Energy 33, 77–87.

Kozempel, M., Cook, R.D., Scullen, J., Annous, B., 2000. Development of a process for detecting nonthermal effects of microwave energy on microorganisms at low temperature. Journal of Food Processing and Preservation 24, 287–301.

Kumar, P., Coronel, P., Simunovic, J., Sandeep, K., 2007. Feasibility of aseptic processing of a low-acid multiphase food product (salsa can queso) using a continuous flow microwave system. Journal of Food Science 72 (3), E121–E 124.

Laguerre, J.-C., Hamoud-Agha, M.M., January 25, 2019. Microwave heating for food preservation and waste exploitation. Available from: https://www.intechopen.com/chapters/65345.

Laguerre, J.C., Pascale, G.W., David, M., et al., 2011. The impact of microwave heating of infant formula model on neo-formed contaminant formation, nutrient degradation and spore destruction. Journal of Food Engineering 107, 208–213.

Lau, M., Tang, J., 2002. Pasteurization of pickled asparagus using 915 MHz microwaves. Journal of Food Engineering 51, 283–290.

Le Bail, A., Koutchma, T., Ramaswamy, H.S., 2000. Modeling of temperature profiles under continuous tube-flow microwave and steam heating conditions. Journal of Food Process Engineering 23 (1), 1–24.

Lin, W., Sawyer, C., 1988. Bacterial survival and thermal responses of beef loaf after microwave processing. Journal of Microwave Power & Electromagnetic Energy 23 (3), 183–194. International Microwave Power Institute.

Lopez-Fandino, R., Villamiel, M., Corzo, N., Olano, A., 1996. Assessment of the thermal treatment of milk during continuous microwave and conventional heating. Journal of Food Protection 59 (8), 889–892.

Luechapattanaporn, K., Wang, Y., Wang, J., Tang, J., Hallberg, L.M., 2004. Microbial safety in radio frequency processing of packaged foods. Journal of Food Science 69 (7), M201–M206.

Maktabi, S., Watson, I., Parton, R., 2011. Synergistic effect of UV, laser and microwave radiation or conventional heating on *E. coli* and on some spoilage and pathogenic bacteria. Innovative Food Science & Emerging Technologies 12, 129–134.

Mallikarjunan, P., Hung, Y.-C., Gundavarapu, S., 1996. Modeling microwave cooking of cocktail shrimp. Journal of Food Process Engineering 19, 97–111.

Morris, C.E., 1991. Food Eng. 63, 98.
Mudgett, R.E., 1989. Microwave food processing. A scientific status summary by the IFT expert panel on food safety and nutrition. Food Technology 43, 117–126.
Nikdel, S., Chen, C., Parish, M., MacKellar, D., Friedrich, L., 1993. Pasteurization of citrus juice with microwave energy in a continuous-flow unit. Journal of Agricultural and Food Chemistry 41, 2116–2119.
Odani, S., Abe, T., Mitsuma, T., 1995. Pasteurization of food by microwave irradiation. Food Hygiene and Safety Science 36 (4), 477–481.
Ortega, Y., Liao, J., 2006. Microwave inactivation of *Cyclspora cayetanensis* sporulation and viability of *Cryptosporidium parvum oocysts*. Journal of Food Protection 69 (8), 1957–1960.
Riva, M., Franzetti, L., Mattioli, A., Galli, A., 1993. Microorganisms lethality during microwave cooking of ground meat 2: effects of power attenuation. Annals of Microbiology Enzymology 43, 297–302.
Ruello, J.H., 1987. Seafood and microwaves: some preliminary observations. Food Technology in Australia 39, 527–530.
Schnepf, M., Barbeau, W.E., 1989. Survival of Salmonella Typhimurium in roasting chickens cooked in a microwave, convection microwave, and a conventional electric oven. Journal of Food Safety 9, 245–252.
Setikaite, I., Koutchma, T., Patazca, E., Parisi, B., 2009. Effects of water activity in model systems on high-pressure inactivation of *Escherichia coli*. Food and Bioprocess Technology 2 (2), 213–221. https://doi.org/10.1007/s11947-008-0069-7.
Shin, J.K., Pyun, Y.R., 1997. Inactivation of lactobacillu plantarum by pulsed-microwave irradiation. Journal of Food Science 62, 163–166.
Stumbo, 1973. Thermobacteriology in Food Processing, 2nd Ed. Academic Press, New York.
Tajchakavit, S., Ramaswamy, H.S., Fustier, 1998. Enhanced destruction of spoilage microorganisms in apple juice during continuous flow microwave heating. Food Research International 31 (10), 713–722.
Tong, C.H., 1996. Effect of Microwaves on biological and chemical systems. Microwave World 17 (4), 14–23.
Welt, B., Tong, C., 1993. Effect of microwave radiation on thiamin degradation kinetics. Journal of Microwave Power & Electromagnetic Energy 28 (4), 187–195.
Welt, B.A., Tong, C.H., Rossen, J.L., Lund, D.B., 1994. Effect of microwave radiation on inactivation of *Clostridium sporogenes* (PA 3670) spores. Applied and Environmental Microbiology 60, 482–488.
Woo, I.S., Rhee, I.K., Park, H.D., 2000. Differential damage in bacterial cells by microwave radiation on the basis of cell wall structure. Applied and Environmental Microbiology 66 (5), 2243–2247.
Yeo, C.B.A., Watson, I.A., Stewart-Tull, D.E.S., Koh, V.H.H., 1999. Heat transfer analysis of *Staphylococcus aureus* on stainless steel with microwave radiation. Journal of Applied Microbiology 87 (3), 396–401.

Further reading

Harlfinger, L., 1992. Microwave sterilization. Food Technology 57–61. December.
Heddleson, R.A., Doores, S., 1994. Factors affecting microwave hearing of foods and microwave induced destruction of food pathogens—a review. Journal of Food Protection 57, 1025–1035.

Hiroshi, F., Hiroshi, U., Yasuo, K., 1992. Kinetics of *Escerichia coli* destruction by microwave irradiation. Applied and Environmental Microbiology 58, 920–924.

Koutchma, T., 1997. Modification of bactericidal effects of microwave heating and hyperthermia by hydrogen peroxide. Journal of Microwave Power & Electromagnetic Energy 32 (4), 205–214.

Ohlsson, T., 1999. Minimal processing of foods with electric heating methods. In: Processing Foods: Quality Optimisation and Process Assessment, pp. 97–105.

Ohlsson, T., Bengtsson, N., 2001. Microwave technology and foods. In: Taylor, S.L. (Ed.), Advances in Food and Nutrition Research. Academic Press, New York, NY, pp. 66–140.

Ramaswamy, H.S., Koutchma, T.N., Tajchakavit, S., 2002. Enhanced thermal effects under microwave heating conditions. In: Welti-Chanes, J., Barbosa–Canovas, G.V. (Eds.), Engineering and Food for the 21st Century. CRC Press, Boca Raton, FL.

Riva, M., Lucisano, M., Galli, M., Armatori, A., 1991. Comparative microbial lethality and thermal damage during microwave and conventional heating in mussels (*Mytilus edulis*). Annals of Microbiology 41 (2), 147–160.

Simunovic, J., Coronel, P.M., 2001. Temperature scan evaluation of continuous microwave processing for pumpable foods and biomaterials. In: Mallikarjunan, P., Barbosa-Canovas, G.V. (Eds.), Seventh Conference of Food Engineering. American Institute of Chemical Engineers, Washington, D.C, pp. 177–183.

Tajchakavit, S., Ramaswamy, H., 1995. Continuous-flow microwave heating of orange juice: evidence of non-thermal effects. Journal of Microwave Power & Electromagnetic Energy 30 (3), 141–148.

Tajchakavit, S., Ramaswamy, H.S., 1996. Thermal vs. microwave inactivation kinetics of pectin methylesterase in orange juice under batch mode heating conditions. Lebensmittel-Wissenschaft und-Technologie 2, 85–93.

Tajchakavit, S., Ramaswamy, H.S., 1997. Continuous-flow microwave inactivation kinetics of pectin methylesterase in orange juice. Journal of Food Processing and Preservation 21, 365–378.

CHAPTER 4

Microwave heating and quality of food

4.1 Introduction

The method of commercial heating operations, reheating and cooking, and delay prior to serving affects the nutritional value of food. The potential of microwave (MW) energy to offer significant reductions in heating time and load during commercial preservation operations such as drying, blanching, pasteurization and sterilization, and food preparation in domestic and commercial ovens makes it attractive to food and food service industry in terms of possibility of achieving better quality and sensory attributes of products. Because microwaves penetrate within the product, heat can be generated in the product's volume and significantly reduce the processing time resulting in greater retention of quality parameters, color, nutrients such as sensitive vitamins, flavor, and aromatic constituents. Microwave-processed foods may also have better texture, taste, and appearance than products processed by conventional methods. Another explanation of achieving better food quality using microwaves is due to fewer loss of moisture contents and consequently better preserving the quality nutrition of foods.

However, the ability of microwave energy to provide better quality than conventional heating is not universally true. When microwave heating occurs less thermally degrading than equivalent conventional process, it is attributed to faster and more uniform heating of the microwaves. When Datta and Hu (1992) evaluated the effects of microwave heating on quality from engineering standpoint based on time-temperature history, they concluded that microwave heating is generally less uniform than a conventional process. Considerable nonuniformity can exit as a result of spatially varying rates of heat generation and consequently overheating some parts of the products. But because the penetration depth of microwaves is under about a few inches or below the surfaces of foods, the uniform heating through overall volume is possible if the sizes of foods are relatively small and the shape of foods is flat with no corners.

If the improvements in quality, nutritional attributes, and taste over conventionally processed products are realized, the premium in product at premium price can be obtainable for such advances. Thanks to these benefits, microwave energy represents today's heat processing method of choice.

Because microwave oven is able to heat up foods quickly, it is well suited to reheat or cook frozen and refrigerated ready-to-eat (RTE) meals. Microwave oven is well suited for cooking the food in small quantities, especially for households though not convenient for mass cooking. In the last decades, cooking with microwave food processing appliances has become the most adaptable method all over the world. Microwave ovens are now used in more than 92% of homes in the United States of America.

Food processors are creating a new generation of microwavable items due to consumers' increasing time pressure, nutritional awareness, and desire for foods that taste and smell like they were cooked in a conventional oven. In many cases, the products' success hinges on a combination of product reformulation, package design, and less processing. This new generation of microwave products includes meals, snacks, and everything in between, from fresh chicken and fish dishes, precooked entrees, and side dishes to grilled cheese sandwiches, biscuits, and pizza.

Among other commercial operations where microwave energy is currently used, the uniqueness of microwave sterilization and pasteurization technology is that it can be applied to both solid and liquid foods as well as complete meals sealed in multicompartment trays. This includes sterilization, pasteurization, and shelf life extension of RTEs. Higher RTE product quality can be achieved due to significantly reduced processing times as compared to conventional thermal processes. The developers claimed that the application of microwaves could reduce the heating time of packaged foods to 1/4–1/10 of time required for conventional methods.

Moreover, microwave preservation of packaged products is possible for different packaging materials such as different polymer plastics, glass, and paper. For microwave sterilization, products have to be packaged in plastic trays or pouches. The ability of plastics to withstand oxygen permeation affects the organoleptic or sensory acceptance of the product during storage. With emerging innovative plastic technologies to the market, the new generations of plastics may increase the shelf life. For example, a two times faster MW pasteurization treatment of pickled asparagus at 915 MHz was published by Lau and Tang (2002). This advantage reduced significantly the thermal degradation of asparagus compared to a conventional treatment in a water bath. Similarly, an acceptable microwave pasteurization of foie gras was reported with a time saving of 50% and better organoleptic qualities compared to a traditional method (Massoubre, 2003).

Furthermore, this technology can be combined with other technologies such as conventional heating using steam, water, or infrared heating for surface cooking. Lau and Tang (2002) successfully applied two successive heating steps, first in water bath and then in 915 MHz microwave oven. Moreover, covering the top one-third of the product glass bottle with aluminum foil eliminated the overheating at the edges. Microwave pasteurization also reduced the cook value for pickled asparagus and reduced textural degradation. On the other hand, the presence of an absorbent medium around the product can reduce the overheating of edges and corners. For example, some authors reported that immersion of the sample in hot water (Guan et al., 2003) or the use of a steam flow into the oven cavity may be used to ensure a safer and better quality of the final product. Microwave-circulated water combination heating system demonstrated a relatively uniform heat distribution within packaged food products.

Like in the studies of microbiological effects of microwave heating, considerable number of the research related to the effects of microwave heating on food quality and composition did not attempt to compare equivalent time-temperature treatments of foods during heating using both heating modes. Based on this, only research that reported sufficient information on heating characteristics was included in this discussion. The objective of this chapter is to present overall effects of microwave heating on food physical properties, quality and nutritional attributes, composition, enzymes in different foods, proteins, lipids, carbohydrates as well as the formation of undesirable chemical compounds such as acrylamides (AAs) and compare those effects with other heating modes in various food processing operations.

4.2 Microwaves heating effects on overall quality of foods

Based on 30-year experience in microwave filed, Decareau (1994) considered that microwave high-temperature short-time (MHTST) sterilization is less successful than conventional retorting of the pouches. The retention of quality parameter of the MHTST process at process temperature of 120–145°C shown in Table 4.1 proves the benefits of such MHTST process. The MHTSP process at 145°C results in 98% retention of thiamin in eel, no texture change in beef, and color in green beans.

Ohlsson (1987) made a comparison of canning, retorting foil pouches, and microwave sterilization of plastic pouches in terms of the cooking value or C-value by integrating the effect of time and temperature on product quality. The results are summarized in Table 4.2 and show the decrease in cooking time a cooking value more than five times when the product was cooked by using microwave energy from 180 min in can retorting process to 28 min for microwave heating of the product in plastic pouch.

Table 4.1 The effect of temperature on various parameters in foil pouch-packed foods.

Parameter	Control unsterilized	Process temperature		
		120	130	145
Thiamin retention in eel (%)	100	17.2	85.3	98.1
Texture in beef (hardness cm/V)	15.2	9.6	13.8	15.1
Viscosity in cream soup (cps)	14.8	20	17.3	15.0
Color in green beans, hue (a/b)	−98.4	−98.4	−72.9	−87.0

Table 4.2 Effect of packaging and process temperature on C-value of a 225-gram pack of solid food.

Package	Dimensions (mm)	Process and process temperature	Cook time (min)	C-value (min)
Can	73 diameter × 49	Retort 120 C	45	180
Foil pouch	120 × 80 × 20	Retort 125 C	13	65
Plastic pouch	120 × 80 × 18	Microwave 128 C	3	28

A comprehensive review covering the effects of microwave heating on physical parameters, quality, and nutritional components of foods has been presented by Cross and Fung (1982). Reviewed data of food composition components included moisture content, color, flavor, animal proteins, nonanimal proteins, carbohydrates, lipids, minerals, and vitamins of fat-soluble and water-soluble varieties. In overall, these authors concluded that no significant nutritional differences exist between foods prepared by conventional and microwave methods.

4.2.1 Overall quality

Bookwalter et al. (1982) from USDA-ARS (Peoria, IL) reported the results of microwave processing of corn-soy—milk (CSM) blends to destroy Salmonellae and evaluate effects on product quality. CSM food blend was developed to supplement the diets of infants and preschoolers. The mixture is composed of partially gelatinized corn meal, defatted, toasted soy flour, and nonfat dry milk. Small amounts of vitamins, minerals, and refined oil were added to provide essential nutrients and improve caloric density. Enriched CSM blends were inoculated with *Salmonella senftenberg* and packed in 22.7 kg bags. The product was heated from 21°C to 56.7 through 82.2°C in 3.9—10 min using a 60 kW (2450 MHz) continuous microwave tunnel. Palletized bags were cooled to 43°C after 9.7 h with moving air. The initial MPN numbers of 4×10^2 CFU/g for *S. senftenberg* were reduced 10^2-10^5 fold after processing at 56.7 through 82.2°C, respectively. The samples were tested for physicochemical parameters, color changes, vitamin A and B, and available lysine. No statistically significant changes in quality were found at process temperatures 67°C and below. Consistency values of CSM decreased after heating at 77.8 and 82.2°C indicating that starch gelatinization occurred. Color changes occurred as well at 77.8 and 82.2°C. However, vitamins A and B showed no significant changes associated with microwave processing treatments. It was concluded that microbial load in CSM can be eliminated by microwave heating with no significant change in product quality.

4.2.2 Moisture content

Moisture content is one of the broadly studied physical parameters of food quality after microwave heating because changes in moisture can be associated with other quality and nutritional losses. Specifically, the effect of microwave cooking and other heating methods on moisture content has been broadly studied in various food categories but mainly in meat and poultry, vegetables, and dairy products. Traditionally, research studies have compared microwave-cooked meat with conventionally cooked meat. Harrison (1980) indicated that early research reports published between 1948 and 1965 indicated that beef, pork, or lamb roasts; chops or patties cooked in microwave ovens usually had higher cooking losses and were less tender, juicy, and flavorful than were comparable cuts cooked by conventional dry heat methods. The author indicated that those studies were not designed for a volumetric microwave heating process, and details of the conditions of microwave heating were not reported or they were reported erroneously. Later studies have shown that microwave-cooked meat can be compared favorably with conventionally cooked meat such as pork and turkey because microwave cooking methods included the use of moist heat and used a low cooking power. For microwave-reheated products, cooking losses were greater and the meat was drier than for conventionally reheated products.

Studies that compared water content of vegetables both fresh and frozen before or after microwave cooking reported few significant data. Differences in water content of vegetables were relatively small. Five cooking methods (two conventional ovens, surface heating, boil-in-bag, and microwave) and their effects on frozen cooked vegetables (green beans, Swiss chard, broccoli, carrots, beets, and potatoes) showed the greatest weight loss in products heated by microwave. Mean losses ranged from 9% of the total weight reported in baked stuffed potatoes to 22% in cut green beans. Also, microwave cooking of vegetables without the addition of boiling tap water resulted in greater weight losses than those incurred through conventional cooking. General l conclusions haven't been made on the basis of existing data, although higher moisture losses by microwave cooking have been indicated in nearly all research to date. The reason of increased losses was associated with a mechanism of microwave heating and greater rise of processing and postprocessing temperature, thus causing more dehydration through evaporation and increased shrinking.

4.2.3 Color

Color and taste are important sensory properties and may be used as a criterion of food quality. However, heating processing can promote reactions that could affect the overall quality of food. The nonuniformity and local overheating of microwave heating can cause not only a severe deterioration of overall quality but also can result in irremediable changes of color where the temperature was the highest (Tang, 2015). These phenomena are mainly observed at the corners and the edges of the product due to wave reflection. As a heating technology, it is hardly avoided that microwave treatment would destroy the pigment that existed in agricultural products. However, compared with other heating methods, microwave treatment can alleviate such a damage to an extent because microwave and radio frequency heating can inhibit the activity of enzyme (polyphenol oxidase [PPO], peroxidase [POD], etc.) effectively.

The comparison effects of four methods of heating such as microwave, ohmic, infrared, and conventional processing on thermal degradation kinetics of color in orange juice were reported by Vikram et al. (2005). The degradation of color as a combination of a- and b-values followed first-order kinetics for all methods of heating. The activation energy values were 14.15 kJ/mK for microwaves, 79.9 kJ/mK for ohmic, and 53.31 kJ/mK for infrared heating. These values fell in the range of reported earlier values of color destruction. However, microwave heating of orange juice led to lower degradation of color compared with other heating methods.

Variability in color and lack of color stability are major problems experienced with microwave-processed fruit products. Ancos et al. (1999) reported the effect of microwave energy on pigment composition and colors of fruit purees such as strawberries, papayas, and kiwi. The samples were treated in 850 W microwave oven at 2450 MHz. Color of purees was evaluated as well as enzyme destruction such as PPO and POD because the activity of these enzymes is attributed to the development of browning, off-flavors, and nutritional damage. Papaya, kiwi, and strawberry purees suffered slight color changes. Contents of carotenoid, chlorophyll, and anthocyanin were evaluated and modifications in composition. The amount of carotenoids were found responsible for overall color in papaya, and the amount of anthocyanin was in charge for color change in strawberry, and chlorophyll degradation changed the color in kiwi. It was found that microwave heating could be an effective treatment to inactivate both enzymes PPO and POD. No temperatures of treatments at various levels of microwave heating were reported in this study.

Brewer and Begum (2003) when studied blanching of selected vegetables reported some darkening due to decreased L-values at 70% and 100%-level of microwave treatment in 700 W oven. Also, hue angle (less green) decreased at 55%, 70%, and 100% power after 1 min of treatments.

The antioxidant ability of lycopene is gaining more interest. Lycopene content (carotenoid responsible for red colors in tomatoes) in raw and thermally processed tomatoes (baked, microwaved, and fried) was investigated by Mayeaux et al. (2006). Samples were baked at 177 and 218°C for 15–45 min, pan-fried at 145°C for 2 min, and microwave cooked in 1000 W oven at 165°C for 1 for 2 min. It was found that 50% of lycopene was degraded at 100°C after 60 min, 125°C after 20 min, and 150°C after less than 10 min. Only 64.1% and 51.5% of lycopene were retained after baking, 64.4% of lycopene remained after microwave heating, and 36.6% and 35.5% of lycopene remained after frying. Microwave heating resulted in the lowest degradation of the lycopene in tomatoes.

4.2.4 Flavor

Clare et al. (2005) focused on the designing and implementing continuous microwave thermal processing of skim fluid milks (white and chocolate) to compare sensory, microbiological, and biochemical parameters with indirect ultra-high temperature (UHT)-treated milks. "No-Bac" Unitherm model from Cherry-Burrell Corporation unit was used for UHT treatment whereas a 60 kW continuous flow microwave heating unit from IMS corporation at 915 MHz was used for microwave heating. Milk was aseptically filled into 250 mL brick-style aseptic cartons. Specifically, bacterial counts, plasmin activity, 5,5′-Dithiobis-(2-nitrobenzoic acid) (DTNB), reactivity protein sulfhydryl groups, sulfhydryl oxidize activity, viscosity, color parameters, and descriptive sensory tests were determined. All test products were aseptically packaged and stored at ambient conditions for 12 months. The results of the tests demonstrated that microwave processing holds much promise for the future milk processing due to better sensory characteristics.

Cremer (1982) found better sensory quality for microwave-heated scrambled egg when compared with eggs prepared in convection oven. However, in case of beef patties, the opposite was observed. The score was higher for appearance, flavor, and general acceptability after heating in the convection oven rather than the microwave oven.

Harrison (1980) reported that microwave-heated meat had more intense meaty-brothy flavor and less stale flavor than did conventionally reheated meat.

4.3 Microwave heating effects on the destruction of vitamins and other nutrients in food

Vitamins can be described as an organic chemical compound (or related set of compounds) of which the organism cannot synthesize in sufficient quantities and must be obtained through the diet. Vitamins have various biochemical functions on human health. Vitamin is a kind of natural antioxidant y and has antitumor activity. It also has a positive effect on the prevention of diseases. In general, vitamins are sensitive to both heat and light. During processing, vitamins are lost due to thermal degradation, photoenzymatic and oxidation reactions.

Cross and Fung (1982) found and analyzed a large amount of available data on the effects of microwave processing on water-soluble (vitamin B complex, ascorbic acid, or vitamin C) and fat-soluble (vitamin A) in buffer systems and foods. No adequate data were reported on the destruction of vitamin A, whereas 11 vitamins of group B complex and vitamin C were extensively studied. They concluded that there are only slight differences between microwave and conventional cooking on vitamin retention in foods.

Thiamine, also known as vitamin B_1, is a vitamin found in whole grains, legumes, and some meats and fish. Thiamine is a sulfur-containing substance easily destroyed by heat or oxidation, particularly in the presence of alkali. Since it is water soluble, it is leached from a product in proportion to the amount of water available, the extent to which it is agitated, and food surface area. The combined effects of temperature, water, and rapid boiling in conventional cooking of vegetables suggested that microwave heating and cooking could be less destructive to thiamine.

The data from various sources proved unsound, as some reports indicated no differences between microwave and conventional heating while other data showed decreased and increased retention of thiamin when treated with microwave energy. There was much variability in the data reported, suggesting that prolonged cooking to reach appropriate end temperatures may be responsible for the apparent thiamine lost in microwave-heated meat.

Meat and meat products represent a good source of dietary thiamine. When cooked beef patties, pork patties, and beef roasts to internal temperatures of approximately 73.5°C in a microwave and conventional oven, there was only an 11% loss of thiamine in beef patties cooked by microwave (Cross and Fung, 1982).

Analysis of thiamine in other studies involving meats (lamb legs and loin chops, beef, top round roasts) revealed similar thiamine retention with microwave and conventional heat treatments. Various precooked and chilled dinners (fried chicken, Salisbury steak, beef, and chicken pies) showed no losses of thiamine when heated in a microwave oven after being stored at −30°C, and some slight increases in thiamine values were reported and concluded that microwave reheating of precooked frozen foods provided the highest thiamine retention. A number of studies strengthened the general perception that cook-chill systems have nutritional benefits, provided that rapid microwave reheating is employed.

Losses of folacin that is another one of the vitamin B group water-soluble vitamin or vitamin B_9 in processing and cooking may range as high as 50%–90% and even 100% when high temperatures and large volumes of water are employed. Although folic acid deficiencies are not usually attributed to inadequate intake, losses are common in infant foods and some reheated precooked foods. Losses variation due to cooking methods were not significant, as retention reported was about 80%. Microwave cooking did not appear more destructive to folacin than conventional methods.

The effects of microwave processing on water-soluble vitamin such as ascorbic acid, niacin (stable vitamin B), thiamin, and riboflavin were determined by Okmen and Bayindirli (1999). They studied kinetics of vitamins microwave degradation first-order reaction. A microwave oven at 700 W equipped with a built-in temperature sensor was used in the study. The reported in this study kinetic parameters such as reaction rate constant k (min^{-1}) at temperatures of 60–90°C, E_a (J/mol, activation energy), z-value (°C, thermal resistance), and Q_{10} value (the ratio of rate constants at temperatures which differ by 10°C) are tabulated in Table 4.3.

The obtained kinetic parameters showed that the most sensitive vitamin was ascorbic acid, followed by thiamin, riboflavin, and niacin. They concluded that microwave heating caused considerably less degradation in water-soluble vitamins when compared to the conventional thermal treatments.

The comparison effects of four methods of heating such as microwave, ohmic, infrared, and conventional processing on thermal degradation kinetics of color and vitamin C in orange juice were reported by Vikram et al. (2005). Freshly squeezed orange juice was used for testing and control. It was found that the destruction of vitamin C was influenced by the method of heating and the temperature of

Table 4.3 Inactivation kinetics parameters of vitamins (Okmen and Bayindirli, 1999).

Vitamins	E_a (J/mol)	z (°C)	Q_{10}	k at 90°C (min^{-1})
Vitamin C	4.58×10^4	65.8	1.41	2.8×10^{-3}
Thiamin	5.47×10^3	83.3	1.32	2.0×10^{-4}
Riboflavin	4.59×10^3	99.7	1.26	1.32×10^{-3}
Niacin	3.65×10^3	125.0	0.86	6.33×10^{-4}

processing. The degradation was highest during microwave heating due to uncontrolled temperature generated during the process. However, higher temperature was employed during microwave heating treatments (100–125°C). Out of four methods, ohmic heating gave the best results facilitating better vitamin retention at temperatures from 50 to 90°C. No attempt to compare equivalency of the treatments was made in spite of the fact that temperature profiles were reported for all four treatments. The reported value of activation energy for vitamin C destruction was 64.8 kJ/mK for MW, 47.27 for ohmic, 39.8 kJ/mK for conventional heating, and 37.12 for infrared heating.

Welt and Tong (1993) compared degradation rates of vitamin B_1 for reactions performed at 100, 110, 120, and 130°C with equivalent thermal histories. Vitamin B_1 is one of the most heat-liable vitamins and has often used as an indicator of the sterilization efficacy of thermal treatments on foods. A special design of microwave heating reactor at 900 W with feedback control system was used in the study, and new methodology to conduct kinetic experiments using microwave energy was introduced. The reported D-value of thiamin in phosphate buffer at 121°C was ??? and Z = 26.6°C. The equation to calculate D-value at any temperature was suggested. No difference in degradation rates of thiamin under microwave heating and conventional heating was found.

Aktas and Ozligen (1992) compared the destruction of *Escherichia coli* bacteria with the degradation of riboflavin (vitamin B_2) in a tubular pasteurization flow reactor. When an inoculum with 10^5 cfu/mL was used, acceptable levels of microbial reduction 0–0.01 was achieved with survival of 60%–75% of initial riboflavin.

The effect of microwave heating on vitamins B_1 and E, and linoleic and linolenic acids and immunoglobulins in human milk was conducted to compare the effects of conventional heating in water bath and MW heating in the oven at 850 W, 2450 MHz at 30% and 100% power settings (Ovesen et al., 1996). It was reported that heating method or power level had no effect on vitamins E and B_1 concentrations heated to 77°C as well as MW heating or water bath did not change the contents of linoleic and linolenic acids. The temperature gradient in nursing bottle under MW heating was approximately 15°C between top and bottom. However, the mean final temperature was rather repeatable. Microwaving at different power levels or conventional heating of breast milk to temperatures approximately of 60°C caused similar destruction of specific immunoglobulin activity. At temperatures of 77°C, almost no activity remained.

Watanbe et al. in 1996 conducted a study to clarify the effects of microwave heating on the loss of vitamin B_{12} in foods, raw beef, pork, and milk. Microwave oven at 500 W at 2450 MHz was used. After samples were treated for 6 min, the loss of vitamin B_{12} was about 30%–40%. The amount of B12 loss in the microwave-treated milk samples did not differ from that in the 30-min boiled milk. The results also indicated that biologically inactive B_{12} degradation products are formed in foods by microwave heating.

While ascorbic acid can exist in a natural or synthetic form, it is found almost exclusively in foods of plant origin such as vegetables and citrus fruits. No appreciable losses in ascorbic acid have been caused by microwave heating of fruits (cooked apples, cranberries, and grapefruit) and vegetables (broccoli, cabbage, beans, tomato, etc.) compared to conventional heating methods. Any increased retention of ascorbic acid in microwave-cooked produce could not be attributed entirely to the features of dielectric heating. A combination of factors including ratio of water-to-vegetable, leaking, cooking length, and variability of cooking loads may account for the variations in the data reported. Spinach and brussels sprouts, cooked, frozen, and kept at $-18°C$ retained 63%–71% of the ascorbic acid of the uncooked vegetables. Reheating by microwave resulted in a 20% further loss of the nutrient. These results compared favorably with losses incurred in domestic cooking.

Koutchma and Schmalts (2002) reported that microwave heating of apple juice to a temperature of 83°C for 50 s (absorbed dose was 1450 J/s) destroyed 20%–28% of the vitamin C. Two microwave ovens (1300 W each) with centrally located helical coils were used to heat apple juice in continuous flow under mixed flow. A water bath was used to heat apple juice flowing through the same coil in conventional water bath heating. Comparison of *E. coli* destruction with the degradation of vitamin C in the range of temperatures tested is given in Table 4.4.

A summary of reported results of microwave heating effects on vitamins A, B_1, B_2, B_{12}, C, and E in groups of different food products and model system is tabulated in Table 4.5.

Brown et al. (2020) compared the effect of two reheating methods on nutrient retention of a frozen single-serve meal that had dual conventional and microwave reheating instructions. The meal has a full range of nutrients that are either thermally labile or thermally stable. Meals were reheated to a minimum of 74°C following the package instructions and analyzed for carbohydrate, protein, fat, sodium, potassium, vitamin A, vitamin C, vitamin E, thiamine, riboflavin, and folic acid. Most of the nutrients studied were statistically unchanged at the 95% confidence level. Potassium, sodium, total fat, total carbohydrates, protein, thiamine, and riboflavin were not affected by the different heat treatments. The changes have been found in the values of other nutrients. Average values of vitamin A and folic acid decreased in both reheated meals compared to the unheated meal; there was no statistical difference between average value after heating in microwave and conventional oven. The average amount of vitamin A decreased by approximately 20%, and the average

Table 4.4 Comparison of *Escherichia coli* inactivation and destruction of vitamin C under microwave heating of apple juice with initial concentration of vitamin C at $C_o = 15$ mg/100 mg.

Temperature (C)	Residence time (s)	Survival ratio N/No	Inactivation (%)	Destruction of vitamin C C/Co (%)
57.2	25	0.45	99.6	0
60	27	0.11	99.9	0
65	29.5	0.00	100.0	0
71	40.6	0.00	100.0	13
75	43.8	0.00	100.0	22
80	51.6	0.00	100.0	28
83	53	0.00	100.0	28

amount of folic acid decreased by approximately 10%. The average amount of vitamin E increased on the order of 20%, and there was no statistical difference between heating modes due to likely release of vitamin E that was previously connected to other compounds. The average amount of vitamin C decreased in the reheated meals. Unlike other nutrients, there was a statistical difference ($P < .05$), although not a clinically meaningful difference (4 mg), between the conventional and microwave heat treatments. The small difference could be due to longer exposure to high temperatures in the conventional oven as compared to the microwave oven. The evaluation of cooking C-value showed that the average C-value for the meals in the conventional oven was 11 and 2.5 min in microwave oven. For this meal, the components were exposed to high temperature four times longer according to the C-value in the conventional oven than in the microwave oven. Data gathered in this study confirmed that microwave reheating does not reduce key nutrients in food beyond conventional oven reheating in single-serve frozen meals. Results of this study supported parity in nutrient retention between reheating methods and suggest that retention of thermally labile nutrients such as vitamin C may be greater after microwave reheating.

4.3.1 Polyphenols

Polyphenols have strong antioxidant activity because they have multiple hydroxyl radicals. It has been reported that phenolic compounds have strong scavenging attraction for peroxide-free radicals. Phenols are found in many green vegetables and fruits.

Effects of microwave and conventional cooking methods were studied on total phenolics and antioxidant activity of pepper, squash, green beans, peas, leek, broccoli, and spinach (Turkmen et al., 2004). Total phenolics content of fresh vegetables ranged from 183.2 to 1344.7 mg/100 g (as gallic acid equivalent) on dry weight basis. Total antioxidant activity ranged from 12.2% to 78.2%. With the exception of

Table 4.5 Summary of reported effects of microwave treatments on essential vitamins in foods.

Vitamin	Product	Microwave treatment	Reported effect	Source
Vitamin A	Corn-soy—milk (CSM)	Microwave (MW) 60 kW 915 MHz	No effect	Bookwalter et al. (1982)
Vitamin B1	Model	700 W, 2450 MHz, 60—90°C	MW inactivation parameters	Okmen and Bayindirli (1999).
	Model	110, 115 and 120°C 2450 MHz. MW reactor	$D_{121} =$ $Z = 26.6°C$	Welt and Tong (1993)
	Human milk	850 W at 2450 MHz at 30% and 100% power settings	No effect at temperatures up to 77°C	Ovesen et al. (1996)
	CSM	60 kW, 915 MHz	No effect	Bookwalter et al. (1982)
Vitamin B2	Model	700 W, 2450 MHz, 60—90°C	MW inactivation parameters	Okmen and Bayindirli (1999)
	Model	Tubular pasteurization reactor	Microbial reduction 0—0.01 was achieved with of 60%—75% destruction of initial riboflavin	Aktas and Ozligen (1992)
Vitamin B12	Raw beef, pork, milk	500 W, 2450 MHz oven, 100°C, 20 min	Loss of about 30%—40% reported. Conversion of vitamin B12 to inactive vit B12 degradation products occurs during MW heating	Watanabe et al. (1998)
Vitamin C	Orange juice	MW at 100—125°C	$E_a = 64.8$ kJ/mK	Vikram et al. (2005)
	Apple juice	MW at 2450 MHz at 1300 W; 1450 J/s 83°C 50 s	20%—28% destruction	Koutchma and Schmalts (2002)
	Model	700 W, 2450 MHz, 60—90°C	MW inactivation parameters; the most sensitive to MW	Okmen and Bayindirli (1999)
	Broccoli, green beans, asparagus	700 W oven, 1—4 min, 50%—100% power levels	Actual concentration increased due to moisture losses, when adjusted moisture losses it decreased	Brewer and Begum (2003)

Table 4.5 Summary of reported effects of microwave treatments on essential vitamins in foods.—cont'd

Vitamin	Product	Microwave treatment	Reported effect	Source
Vitamin E	Egg yolk	1500 W, MW oven	Reduced by 50% traditional cooking and MW	Murcia et al. (1999)
	Human milk	850 W at 2450 MHz at 30% and 100% power settings	No effect at temperatures up to 77°C	Ovesen et al. (1996)
	Soya bean oil	500 W oven, 2450 MHz, roasted for 4–20 min	Remained >80% in soaked beans after 20 min of roasting	Yoshida and Takagi (1996)

spinach, cooking affected total phenolics content significantly ($P < .05$). After cooking, total antioxidant activity increased or remained unchanged depending on the type of vegetable but not on the type of cooking. The effect of various cooking methods on total phenolics was significant ($P < .05$) only for pepper, peas, and broccoli. Microwaving and steaming caused a loss in phenolic content in squash, peas, and leeks, but not in spinach, peppers, broccoli, or green beans. The authors concluded that moderate heat treatment might have been considered a useful tool in improving health properties of some vegetables.

In microwave-assisted extraction, microwaves tend to increase the total phenolic content significantly. On one hand, rapid heating induced by microwave radiation caused the plants to release polyphenols to resist external interference such as air oxidation (Barcia et al., 2015). On the other hand, the combination of microwave energy and moisture penetrated the plant matrix, promoted the dissolution of polyphenols in cell tissue, and increased solubility. Increasing extraction time and power levels led to the increase of total phenolic content in the extract. However, for heat-sensitive polyphenols, such as proanthocyanidins, higher temperature accelerated decomposition and destruction. When using microwaves in extraction, attention should be paid to the decomposition temperature of the extracted material to avoid damaging the compound by extending the extraction time, and direct extraction of heat-sensitive substances is not suitable for the microwave method (Hu et al., 2021).

4.4 Microwave heating effects on lipids, proteins, and carbohydrates in foods

According to Cross and Fung (1982) in overall, the effects of microwaves on protein, lipid, minerals, and carbohydrates in foods appeared to be minimal. Variability in procedures and products prevents any additional conclusions from the research data. However, it was observed that carbohydrate, lipid, and protein are sensitive to microwave heating.

4.4.1 Edible oils and fats

Lipid is a group of compounds that are generally soluble in organic solvents and largely insoluble in water. Lipids may be broadly defined as hydrophobic or amphiphilic small molecules. Lipid is a strong polarity molecule for analyzing the structure of lipids, and lipids always showed high dielectric constant and loss factor at a certain frequency. It was reported that lipids can be oxidized and degraded during heating. Free radicals are produced during microwave heating, which makes lipids more susceptible to oxidation under microwave heating conditions than conventional heating.

Research has been done on microwave heating effects on lipid composition in foods including the degree of lipid oxidation, total fat losses, and fatty acid composition. The comparison of effects of microwave and conventional heating on thermooxidative degradation of edible fats and oils such as sunflower oil, high oleic sunflower oil, virgin olive oil, olive oil, and lard was carried by Albi et al. (1997) in well-controlled treatments in 1 kW microwave oven, conventional oven, and exposed to microwaves without heating. Degradation was quantified by means of measuring of acid value, peroxide value, and a-tocopherols. Analysis showed greater alterations in microwave-treated samples than in the samples treated by conventional methods. Microwave energy treatment without heating produced no oil alterations. Nezihe et al. (2011) found out that the fatty acid composition after microwave heating determined by gas chromatography analysis showed the increased content of unsaturated fatty acid.

The fatty acid composition of soy beans was not altered by microwave heating to 81°C (Hafez et al., 1985). Some researchers agreed that the lipids and lipid-containing foods were sensitive to microwave heating, which made it easy to be oxidized and degraded.

Rancidity of cottonseed oil under conditions of microwave heating occurred fast, and the content of peroxide and secondary oxidation products, and levels of peroxide in the product increased, and the degree of peroxide reactions positively correlated with microwave power levels. The reaction mechanism of peroxide is related to reactive-free radical methylene in lipids, produce the free radicals which easily react with oxygen to produce peroxide. Microwave radiation and high temperatures accelerate the formation of these free radicals. Lipid can be degraded to secondary oxidation products after peroxidation, which is the main factor in oil rancidity (Farag et al., 1992).

Overall, any effects of microwaves on the lipid fraction in foods appeared to be minor—only small differences existed in total fat and fatty acid composition. No general statement can be made regarding total fat changes in microwave versus conventionally prepared food products from the published research and have to be examined for specific processing conditions, food products, and packaging.

4.4.2 Proteins

Protein is a complex substance that is presented in all living organisms. Proteins play critical roles in nearly all biological processes, including catalyzing metabolic reactions and DNA replication, and transporting molecules from one location to another. Most proteins fold into unique three-dimensional structures. Denaturation of proteins destroys their normal physiological functions, damages cells, and causes disease. In the process of heating, the protein is prone to denaturation and produces carcinogens. Several studies have demonstrated that exposure to microwave radiation can cause nerve damage in humans, as well as oxidative stress and inflammation.

The effects of microwave heating on proteins of animal and plant origin were studied. Proteins and peptides have high values of dielectric constant, and their structure and activity can be affected by microwave heating. The changes in protein denaturation were determined through measurements of total proteins, nitrogen distribution between sarcoplasmic and crude myofibrillar fractions, values for insoluble protein, free amino acids content, and proteins digestibility. It is widely drawing the attentions that microwave treatment has significant effect on protein degradation and accelerating reaction. However, the nutritional effects of microwaves on animal proteins in meat, poultry, and fish appear minor.

The studies of effects on plant proteins included potatoes, beans, soy, peas, and other legumes. Microwave radiation appears to have little if any adverse effects on plant proteins; in fact, one study was found that microwave radiation destroyed harmful proteins and improved some properties of proteins. The effects of exposure to heat and radiation and the subsequent formation of free radicals damage the intramolecular forces, causing gradual exposure of tryptophan, initially trapped in molecules, it breaks free of intramolecular hydrogen bonds causing the orderly arrangement of atoms in proteins to become disordered. Under the influence of electromagnetic fields, polar radicals in proteins rotate and collide, and the frequency and intensity of interactions between hydrophilic groups on the surface of proteins and the water molecules increase which promotes hydration. At the same time, the destruction of the intramolecular forces in protein molecules promoted aggregation of loose proteins (Hu et al., 2021).

Conducted experiments were to determine in vivo protein digestibility and metabolizable nitrogen of soy proteins using male rates. Microwave heating of soybeans in the 650 W oven operated at the full power for 9 min decreased protein solubility from 80% to 17%, from 81% to 18%, and from 72% to 16% when deionized H_2O, 0.6 N NaCl, and 0.4 N $CaCl_2$ were used as solvents. Majority of studies indicated favorable changes in the protein fraction of plant food exposed to microwave compared with conventional cooking. While no deleterious effects on nutritive value were reported in the literature, susceptibility to heat processing has been observed with protein foods containing significant amounts of carbohydrates. It is generally accepted that interactions between functional groups with the protein chain or between the protein chain and other food constituents during heating lead to cross linkages that can be resistant to the normal digestive processes.

After heating, the level of thrombin, a protein that can be fatal and is commonly found in legumes, decreased, and the destruction of trypsin inhibitors was greater in wet beans. Therefore, the use of microwave to heating bean seeds not only reduces toxic proteins found in these seeds but also improves digestibility. Hydration of proteins also facilitates their emulsification, but aggregation of proteins and exposure to hydrophobic radicals reduce their solubility. The effect of microwaves on secondary structures found in proteins was to increase surface area, thus microwave technology can be used to pretreat proteins through enzymatic hydrolysis (Hu et al., 2021). In summary, protein will be denatured after microwave treatment, breaking down harmful proteins effectively.

4.4.3 Carbohydrates

Carbohydrates are common in human diets, such as potatoes, wheat, maize, rice, and cassava. The most common carbohydrates existing in the plant tissues are starch and saccharides, which include fructopyranose, saccharose, amylaceum, etc. An extensive literature search revealed no report on the effects of microwaves on the carbohydrate fraction in foods. Microwave treatment can change the structure and crystallinity of starch (because of molecular vibration), and therefore the characteristics of starch such as polarity, free energy, viscosity, gelatinization, molecular weight, particle size, etc. (Szepes et al., 2005). It can be observed that the crystal structure of potato starch changed from type B to type after being processed by microwave heating. The impact of microwave heating on the physicochemical properties of a starch—water model system was significantly different from conduction-heated gels in all parameters measured, while the lack of granule swelling and the resulting soft gel were two key observations. Many studies reported that the microwave treatment can accelerate the synthesis/hydrolysis rate of starch significantly, and meanwhile some reaction can be catalyzed by microwave treatment. The change of carbohydrates' structure is always accomplished with the changes of physicochemical properties. More research has to be done in this area (Jiang et al., 2017).

4.4.4 Polysaccharides

Polysaccharides are bioactive ingredients that show significant biological functions with many protective properties. Polysaccharides are antioxidant with immunomodulatory factors, antitumor formation, which can protect against hypoglycemic, hypolipidemic, and gastrointestinal conditions.

After conventional drying and extraction, the structure, physical and chemical properties of polysaccharides in natural products can be changed. Also, the distinctive porous structure formed by microwave heating can modify properties of the polysaccharides. Multiple studies have shown that microwaves influence bioactivity of polysaccharides, and their influence is related to their structure. After microwave treatment, the viscosity, enzyme activity, content, and molecular weight of the polysaccharides may change, which affects their functional characteristics. In the process of microwave treatment, polysaccharides are susceptible because of their intensive polarity. The porous structure formed by microwave heating has a beneficial effect on the hypoglycemia activity of polysaccharides. The breakage of chemical bonds caused by microwaves can release smaller molecule of sugars and promote antioxidant activity. However, the reduced viscosity may affect its adhesion and affect the biological activity. The antitumor activity of polysaccharides can be lower after microwave treatment because of their lower viscosity.

4.4.5 Starch

Starch is a polymeric carbohydrate consisting of numerous glucose units joined by glycosidic bonds. This polysaccharide is produced by most green plants for energy storage. Compared with the biological activity of polysaccharides, more attention is focused on the starch's digestibility. Microwave treatment can result in slower digestible of starch, which is suitable for patients with diabetes. It was found that the structural changes of starch during microwave treatment could lead to gelatinization of starch and cause modifications in physical and chemical properties, which are the factors leading to the changes of digestibility and stability. Microwave heating of polar molecules caused the surface of starch particles to crack and shrink. Polar radicals such as hydroxyls, carboxyls, and water molecules in starch vibrated rapidly, and as a result, destroyed the double helix structure of starch, molecules of chemical bonds also break and enhance intermolecular force. Additionally, the double helix structure of starch can be destroyed due to rapid vibration of the molecule. Unlike the traditional heating, the selective effect of microwave heating on polar bonds can form the α-(1,6) glycosidic bond in the outer chain that can be damaged, and the broken starch chain is rearranged. During such rearrangement, starch molecules become polarized and result in increase of the dielectric constant parameter. The broken amylopectin is rearranged into amylose. It appeared that after microwave treatment, the starch not only expanded but also formed pores. The main cause of starch crystallization is attributed to amylopectin. After microwave treatment, the crystallinity of starch molecules generally decreases, which indicates on reduction of amylopectin. The breakage of molecular chains also contributed to the degradations of starches to form monosaccharides and increased acidity. Such rearrangement of molecular chains of starch improved the combination between starch and water molecules, causing expansion and better gelatinization of the starch. Gelatinization of starches is usually accompanied by an increase in viscosity. When starch molecules vibrate during microwave heating, particles of starch rupture and amylopectin structure rearranges causing viscosity increase. When these changes in structure occur, the digestibility of the starches increases because the internal structure of the starch has been damaged. Appropriate microwave power levels and times of exposure should be adjusted according to the level of digestibility of the starch after microwaves treatment.

In summary, microwaves can directly or indirectly induce a series of changes in the morphology and internal structures of starch granules, closely related to the dielectric properties of the system and microwave input energy. The future research should be aimed on manipulating the structural changes of microwave-treated starch systems from the perspective of obtaining microwave-treated starch-based food with desirable properties.

4.4.6 Microwave heating for developing foods with low glycemic index

The glycemic index (GI) of foods can be reduced by increasing resistant starch (RS) or slowly digestible starch (SDS). Rapidly digestible starch promotes metabolic syndrome, and thus influence insulin resistance, obesity, and type 2 diabetes. The benefit of SDS is a moderate impact on the GI. RS within a calorie-controlled diet is beneficial in protecting against metabolic syndrome and colon cancer. Structural changes to starch have been reported to be one of the approaches adopted in reducing the GI of food. The use of microwave technology to produce RS has been sufficiently reported in comparison to the other technologies.

According to the report of Mapengo and Emmambux (2022), a moisture content (30% and 90%), microwave power level between 300 and 1000 W, and a treatment period of 0.5–10 min have been shown to increase the amount of RS for various starchy foods. The high moisture and microwave power combination promotes starch gelatinization and retrogradation during cooling. Li et al. (2020) and Canas et al. (2020) reported an increase in RS and SDS in rice starch and pasta after microwave cooking combined with cold storage for a period of about 24 h. Cold storage promoted amylose recrystallization after gelatinization thereby resulting in RS type 3. Microwave heating has been reported to decrease the branching degree of amylopectin, resulting in the retrogradation of linear chains, and further promoting the formation of RS during cooling (Zeng et al., 2016). Also, it was reported that extended microwave treatment could promote gelatinization and resulted in a greater degree of starch degradation by more fully destroying the starch crystalline structure, thus making enzyme easier accessible to the alpha glycosidic bonds. Clearly, reported research demonstrated that the microwave technology can be used to produce rapidly digestible starch or RS depending on the processing conditions and moisture level.

4.4.7 Minerals

Limited data are available concerning specific effects of microwaves on mineral composition. Overall, it has been assumed that the mineral composition of foods of animal and plan origin is not altered. However, there are mineral determinations for meat drippings that showed significant differences for sodium, chloride, phosphorus, and iron content of drippings from oven-roasted pork compared to microwave-heated pork; this trend was evidenced in phosphorus and iron content from beef and lamb drippings as well suggested that this greater concentration of minerals in the drippings from the conventionally cooked roasts could be related to the lower moisture content of the drippings (Cross and Fung, 1982).

4.5 Microwaves heating and enzymes destruction in foods

A number of studies were conducted to look at the destruction of thermal stable enzyme under microwave heating that often serves as an indication of thermal treatments during blanching or pasteurization.

Henderson et al. (1975) reported the results of horseradish POD destruction. The samples were subjected to 2450 MHz continuous microwave radiation with a power from 5 to 600 W over an absorbed power density from 62.5 to 375 W/cm^3 for time periods of 5, 10, 20, 30, and 40 min at 25°C during exposure. The significant inactivation of the enzyme was found only for absorbed power densities greater than 125 W/cm^3 for 20 min. It was concluded that microwave heating did not have an effect on enzyme activity aside from thermal effects.

Brewer and Begum (2003) when studied blanching of selected vegetables reported reduced POD activity after one minute of microwave treatments in 700 W oven. Ramaswamy and Fakhouri (1998) found microwave and thermal inactivation of POD in frozen vegetables comparable.

Browning reactions in fruit and vegetables are due to the activity of PPO. A microwave applicator at 113 W in a waveguide at 2450 MHz was designed and used to study inactivation kinetics of mushroom PPO under microwave and conventional heating (Rodrieguez-Lopez et al., 1999). It was found that microwave and conventional heating produced different enzyme intermediates with different stability and kinetics parameters. It was reported that considerable time can be saved during microwave inactivation resulting in greater profitability and enhanced quality.

Kamat and Laskey (1970) studied the effect microwave radiation at 2450 MHz on aqueous solutions of *Bacillus subtilis* α-amylase (BAA). They were unable to demonstrate any enzyme inactivation in samples irradiated at an incident power density of 106.8 mW/cm^2 for 15 min in which temperature was maintained in the range of 16–38°C. The enzyme was significantly inactivated (62.78%) when the temperature of the solutions was increased to 63.7 ± 11.1°C. Further exposure of enzyme solution to temperature of 72.0°C for 15 min resulted in 96.5% inactivation of α-amylase.

Tong et al. (2000) studied kinetic parameters of BAA destruction when used as a biological time-temperature integrator (TTI) to assess the relative effectiveness of continuous-flow microwave heating for enzyme inactivation under changing fluid temperatures and flow rates. A continuous-flow microwave heating system was set up using two microwave ovens (Sharp R-114AWC, Memphis Tennessee; Gold Star MS-133XC, LG Electronics Inc, Canada; 1000 W nominal output power capacity each, 2450 MHz). Mean bulk temperatures of the fluid were continuously recorded by copper-constantan thermocouples at the inlet and outlets of the ovens. Two helical coils made out of glass tubing (internal diameter 8.9 mm and length 2.0 m) were positioned at the center of the microwave cavities for continuous heating test samples. Heating characteristics as time-temperature curves, temperature rise versus flow rate/residence time, heating rates, microwave absorbed power, Reynolds and Dean numbers for water and enzymes solutions were determined, and the results are summarized in Table 4.6.

Table 4.6 Summary of operating temperatures, flow rates, heating rates, and absorbed power.

Exit temperature (°C)	Flow rate (mL/s)	Residence time (s)	Heating rate (°C/s)	Power absorbed (W)	Efficiency (%)	Re/De numbers
95	4.17	64.8	1.09	1228.2	0.61	600/178
88	5.00	54.0	1.18	1327.3	0.66	720/214
81	5.50	49.1	1.14	1287.3	0.64	792/236
75	6.67	40.5	1.23	1395.7	0.69	960/286
70	7.50	36.0	1.25	1413.1	0.71	1080/322
66	8.33	32.4	1.27	1441.3	0.72	1200/357

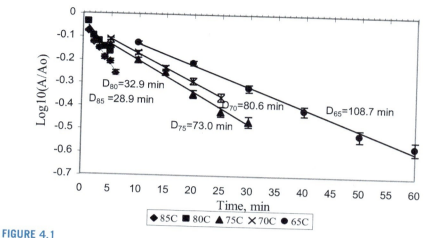

FIGURE 4.1

Thermal inactivation curve of *Bacillus subtilis* a-amylase at 0.1 mL/L (uncorrected).

Isothermal inactivation curves and kinetic parameters of α-amylase in phosphate buffer (pH 6.9) estimated using a first-order reaction are shown in Fig. 4.1. Using decimal reduction times in the range of temperatures from 65 to 85°C, the z-value of heat inactivation of BAA was obtained equal to 32.5°C under conventional heating mode.

The relatively high magnitude of D- and z-values under isothermal heating in buffer of pH 6.9 resulted in low destruction of the enzyme under continuous-flow microwave heating. No destruction was observed in the range of temperatures from 70 to 90°C. Increasing exit temperature of the enzyme solution up to 95°C allowed to achieve only 0.07% of destruction (Table 4.2). The pH dependence of the decimal reduction time of α-amylase was studied by heating test samples in a water bath at pH 6.0. The uncorrected D-values at 80°C were 32.9 min at pH 6.9, 5.8 min at pH 6.5, and 0.73 min at pH 6.0, and thus the pH modification offered a potential approach to change thermal resistance of this enzyme. To validate the technique, the destruction of α-amylase was evaluated under nonisothermal microwave continuous-flow conditions using phosphate buffer of pH 6.0 as a heating media. The experimental results at selected temperatures and flow rates are also summarized in Table 4.7. The obtained data showed the sensitivity of α-amylase to environmental change of pH in the range of 70–90°C and demonstrated the potential of TTI in the context of current application.

Kermasha et al. (1993a,b) analyzed the inactivation of wheat germ lipase and soybean lipoxygenase at various temperatures using conventional and microwave batch heating and found higher enzyme destruction rates under microwave heating conditions. Lipoxygenase inactivation was evaluated using first-order reaction kinetics. The activation energy E_a were 119 and 180 kJ/mol, respectively, for the conventional and microwave heating modes.

Table 4.7 Destruction of α-amylase under nonisothermal continuous-flow microwave conditions.

pH of heating medium	Exit temperature (°C)	Flow rate (mL/s)	Mean residence time (s)	Residual activity (A/A$_o$)
6.9	95.0	4.17	64.80	0.93
6.9	88.1	5.00	54.00	1.00
6.0	70.2	7.67	35.22	0.84
6.0	75.0	6.83	39.51	0.79
6.0	80.1	6.17	43.78	0.64
6.0	86.1	5.50	49.09	0.33

Wang and Toledo (1987) used a microwave oven to inactivate soybean lipoxygenase as a function of moisture content. Complete inactivation of lipoxygenase in soybeans with moisture content between 27% and 57% was reached after 210 s of microwave heating. They also concluded that the time of microwave heating is shorter in comparison with conventional heating. They also postulated that microwave treatment of soybeans at their natural moisture content could provide suitable material for soybean processing.

The studies of destruction pectin methylesterase (PME) by Tajchakavit and Ramaswamy (1995, 1996, 1997) were conducted under continuous-flow and batch heating modes. Thermal resistance of pectin PME implicated in the loss of cloudiness of citrus beverages has been recognized to be greater than that of common bacteria and yeast in citrus juices, and PME activity has been used to determine the adequacy of pasteurization.

Conventional heat treatment of orange juice (2 mL) was performed in water bath at 60, 70, 80, 85, and 90°C at pH 3.7. For batch mode microwave heating, a 700 W microwave oven was used to heat test samples in a 100 mL cylindrical glass container to yield target temperatures 50, 55, 60, and 65°C. To obtain the final bulk temperature, the test sample was mixed immediately after heating and temperature was measured in a well-insulated container to prevent heat losses during temperature measurement. For continuous-flow microwave heating, the setup described earlier was used. Different lengths of heating tubes were employed to get different effective times at the exit temperatures at 55, 60, 65, and 70°C.

The kinetics parameters (D- and z-values) obtained from the slopes of linear sections of the different curves are included in Table 4.8 for thermal as well as microwave heating conditions.

Thus, the PME destruction in orange juice was significantly faster in the microwave heating mode than in conventional thermal heating mode. Inactivation kinetics was also dependent on pH and soluble solids content of orange juice. PME was more sensitive to thermal treatment inactivation at lower pH and higher soluble solids concentration.

Table 4.8 D- and z-values under thermal and microwave destruction of PME enzyme in orange juice.

Temperature (°C)	Thermal D-value (s)	Microwave batch D-value (s)	Microwave continuous D-value (s)
50.3	–	40.5	
55.3	–	11.7	38.5
60.0	154	7.37	12.4
64.9	–	2.96	3.98
70	37.2		1.32
80	8.45		
85	6.53		
90	2.85		
z value (°C)	17.6	13.4	10.2

The brief summary of the data published in the reviewed research papers on microwave heating effects on the destruction of enzymes is given in Table 4.9 indicating disparities in reported effects such as faster inactivation of PME enzyme, differences in activation energy, and no effects on POD enzymes in some foods.

4.6 Effects of microwave heating on food chemistry
4.6.1 Chemical reactions

Flavors and colors generated as a result of the Maillard reaction in microwave heating was discussed by Yaylyan and Roberts (2001). Maillard reaction is a complex series of consecutive reactions and depends on temperature, water content, pH, and heating time. The products of this reaction result in color, volatile aroma compounds, and nonvolatile products. The specific features of microwave heating may affect chemical reaction and may enhance or slow them down. Superheating of solvents (increase temperature of liquid above their boiling point) is one phenomenon that accompanies microwave heating and helps accelerate chemical reactions. At the molecular level, the interaction of microwaves with the food matrix and specific compounds will affect the chemical reaction in general. The most of model studies reported in literature on this matter are reviewed in this chapter by Yaylyan and references can be found there.

Browning and good texture at a fixed moisture level are essential in bread baking process. Conventional baking using hot air provides suitable color and texture. In microwave baking, sufficient brown color on the surface of breads and crust formation were not possible due to its volumetric nature and cold air surrounding the food product. Water that evaporates from food surface is condensed in contact with cold

Table 4.9 Summary of reviewed research papers on microwave destruction of enzymes.

Enzyme	Product	Microwave treatment	Reported effect	Source
Pectin methylesterase (PME)	Orange juice	700 W, microwave (MW) oven 50–90°C. Compared with conventional heating	PME destruction in orange juice was significantly faster in the MW heating mode than in conventional thermal heating mode.	Tajchakavit and Ramaswamy (1995, 1996, 1997)
Lipoxygenase	Wheat Soybean	700 W; 2450 MHz MW oven	The activation energy E_a was 119 and 180 kJ/mol for the conventional and MW heating modes. Time to completely inactivate enzyme by MW heating is shorter than conventional heating.	Kermasha et al. (1993a,b) Wang and Toledo (1987)
α-amylase	Aqueous solutions Phosphate buffer	Two 1000 W MW ovens with a coil	62.78% inactivation at T = 63.7°C. T = 72.0°C for 15 min resulted in 96.5% inactivation. Sensitive to change of pH in the range of 70–90°C. MW heating to 95°C for 65 s allowed to achieve 0.07% destruction.	Kamat and Laskey (1970) Tong et al. (2000)
Peroxidase (POD)	Model solutions Blanching vegetables Frozen vegetables	Continuous MW heating with a power from 5 to 600 W at 2450 MHz 700 W MW oven 2450 MHz	MW did not have effect on enzyme activity aside from thermal effects. Reduced POD activity after 1 min of MW treatments. Comparable MW and thermal inactivation of POD in frozen vegetables.	Henderson et al. (1975) Brewer and Begum (2003) Ramaswamy and Fakhouri (1998)
Polyphenoloxidase (PPO)	Mushrooms	MW applicator at 113 W in a waveguide at 2450 MHz	MW and conventional heating produced different enzyme intermediates with different stability and kinetics parameters.	Rodrieguez-Lopez (1999)

air and results in the lack of crispness of bread product. Susceptors, microwave absorbing materials, are commonly placed at the bottom of the sample to provide crust formation and surface browning of the food product. Browning was not observed without the help of susceptors. Lightness values were found to decrease with baking time and temperature which indicated that the color of the sample became darker. The results showed that zero-order kinetics can explain the change in lightness using microwave heating.

A crisp crust and brown color can be obtained with microwave along with jet impingement. Baking of bread using different heating modes such as jet impingement, microwave plus jet impingement, and microwave plus infrared were investigated by Datta et al. (2007). Jet impingement baking is carried out by applying high speed convection with the help of commercial electrical oven. The air jets were introduced from top to bottom at a velocity of 10 m/s. Microwave–impingement combination baking is achieved by combining high-speed convection heat with microwaves using the same JET oven. Here, microwaves were introduced from the top and the air jets were introduced from both top and bottom at a velocity of 10 m/s. Microwave infrared baking combines both microwave and infrared heating. In the combination oven, halogen lamps are installed at the top (oven ceiling) and bottom (oven floor) and a rotary table was provided to improve heating uniformity of the samples. The results showed that the temperatures were higher in microwave-infrared combination, intermediate in microwave–impingement combination, and lower in JET oven. Maximum moisture was retained in breads baked in JET compared to those baked in other combined modes. Since microwave plus infrared did not develop good, rigid outer crust, the temperature at the surface was lower and a lot of moisture had escaped. This results in the reduction of final volume of the bread.

Based on these reports, it is concluded that microwave heating alone or in combination with other modes of heating such as hot air or infrared radiation for baking process do not provide better end-product qualities compared to that of conventional baking.

The possible toxic compounds formed during the thermal processing of food products include, among others, heterocyclic aromatic amines, nitrosamines, polycyclic aromatic hydrocarbons, 5-hydroxymethylfurfural, furan, and AA. One of the challenges is to minimize the formation of this AA during industrial food processing as well as food preparation by consumers. Home-cooking choices could have a substantial impact on the level of AA.

4.6.2 Acrylamide formation

AA is a substance that has recently been discovered in food. The major pathway of AA formation seems to be the Maillard reaction, in particular, in the presence of the amino acid asparagine, which directly delivers the backbone of the AA molecule. Relatively high amounts of AA, a potential carcinogen to humans were detected in thermally processed carbohydrate-rich foods, for example, potatoes.

One of the factors affecting its formation is the heat treatment method. Available studies demonstrate differences in the mechanisms of microwave and conventional heating. These differences may be beneficial or detrimental depending on different processes. The formation of AA upon treatments with microwave and conventional heating, boiling or frying was investigated in ASn/Fru and Asn/Glc model systems and potato chips (Yuan et al., 2007). The effect of pH was found complex in both systems during microwaving at 600 W and boiling at 120°C. It was found that regardless of pH and heating methods, AA content increased with treatment time. Microwave heating induced more AA formation and facilitated its formation compared to boiling method at identical pH. It was observed for both model systems and potato chips. Yuan et al. (2007) suggested that microwave power for carbohydrate-rich food processing should be used at the lowest possible level. They reported that, in comparison to conventional heating, such as boiling or frying, strong microwave treatment is more favorable for the AA formation in both asparagine/fructose and asparagine/glucose model systems and in potato chips.

There are other mechanisms of microwaves' influence on food ingredients. Rydberg et al. (2005) studied the effect of microwave heating on mashed potato and showed that the formation of AA progresses with the pyrolysis of the sample. The increase of the AA content was 140 times higher after prolonging the heating time from 100 to 150 s. Microwave heating gave rise to uncontrollable variations, mainly due to inhomogeneous heating resulting in local pyrolysis.

Michalak et al. (2020) concluded based on the published research that microwave heating at a high-power level can cause greater AA formation in products than conventional food heat treatment. The higher content of AA in microwave-heated foods may be due to differences in its formation during microwave heating and conventional methods. It was concluded that the formation of AA in microwave-heated food, as in the case of conventional heating, depends on the process parameters and the properties of the processed products. At the same time, short exposure to microwaves (during blanching and thawing) at low power may even limit the formation of AA during the final heat treatment. Considering the possible harmful effects of microwave heating on food quality (e.g., intensive formation of AA), further research in this direction is essential.

4.6.3 Thiobarbituric acid values

The autoxidation is used to describe the chemical degradative reactions which cause oxidation rancidity. The rate of autoxidation is enhanced by the presence of prooxidants, such as microwave heating. Thiobarbituric acid (TBA) values have been used as a measure of lipid oxidation for a variety of foods.

Microwave-reheated turkey breast muscle was analyzed using the TBA test to determine the extent of oxidation because this is an indirect measure of reactive oxidative species. TBA values were compared to those of raw, freshly cooked, and conventionally reheated turkey breasts (Bowers, 1972). TBA values for all samples were fairly low (1.60 for freshly cooked, 1.54 for microwave reheated, and 1.71

for conventionally reheated samples) and did not differ statistically for the three heated samples. However, they were higher than values reported for the raw muscle tissue (0.33).

4.7 Conclusions

Because conventional methods of heat treatments for food preservation can negatively impact end-product quality, the aim of modern advanced food processing technologies is to improve quality, and enhance quality, organoleptic and healthier nutritional properties combined with better food safety. Furthermore, conventional methods are not optimized for solid foods because of the conduction mode of heat transfer from the surface to the cold point that is often located at the center of the product. Microwave heating obviously has the advantages to overcome the limitations of slow thermal diffusion due to its volumetric nature and has experienced a growing industrial demand due to its flexible and rapid heating performance. Microwave heating is successfully commercialized for food drying, defrosting and preservation in commercial processing operating, and reheating and cooking in domestic ovens.

Microwave energy represents an advanced method of modern cookery, especially with respect to efficiency of operation, energy savings, and reduced cooking time. Current microwave oven sales are indicative of consumer demand and the growing popularity of microwave cooking.

However, this process is still relatively poorly controlled because of complex interactions between foods components and microwave energy. Furthermore, the heating heterogeneity is the major drawback of this technology and results in overheating and negative effects on food quality.

The evaluation of existing data shows that under equivalent conditions, no significant nutritional differences exist between foods prepared by conventional and microwave methods. Any differences reported in the literature are minimal. With further modifications and improvements in magnetrons and oven design, current microwave ovens will heat foods more uniformly and reliable, thus leading to a better final product during preheating, coking, or defrosting.

It is controversial that heat effect is the only fact to affect the characteristics of components inside foods during microwave treatment, especially the nutrition loss. However, by analyzing the related articles, it was observed that carbohydrate, lipid, and protein are sensitive to microwave treatment. The mechanism of component changes and interactions between different components during microwave heating is not well defined.

The future success of the microwave systems depends upon the continued improvement of engineering and research advances in areas of products formulation, packaging, and consumer education. The microwave industry has maximized the popularity and interest in microwave cooking and promoted new ideas to cross the once-rigid technical boundaries.

An industrial effort should be made by food processors to educate consumers by providing detailed cooking instructions on packages, standardizing basic recipes, and improving methods for cooking complete meals.

The efficiency of microwave-based processes is depending on parameters related to the intrinsic characteristic of the treated food, the design of the equipment, and the process itself that can interfere with the heating rate. Therefore, understanding these parameters is essential to optimize the process, and thus ensure high heating rates that will increase energy efficiency and avoid economic losses, as well as minimize thermal damage while preserving the nutritional, functional, and sensory characteristics of the products.

Further research on microwave effects on quality and nutrients is needed, with research emphasis on moisture, protein, carbohydrate, vitamin retention, and chemical reactions. A collaborative effort by microwave researchers to standardize experimental procedure, measure, and report heating characteristics of the systems and of the products would be highly desirable to minimize research variability. The development of microwave cooking techniques can be pursued to optimize food quality. Microwave penetration, shapes, and geometry of packaged products must be taken into account to avoid overheating.

To get the products with improved quality after microwave heating, equipment design and operation should meet the following requirements: electromagnetic field nonuniform distribution, corners, and edges effects should be avoided by optimizing the design of microwave systems and changing the shape of materials; heating at reduced power level for long duration is good for quality; movement of materials ensures product heating uniformity; and accurate temperature monitoring is important for high-quality products' processing. Traditional temperature measurements (thermocouples or infrared temperature meter) are not available under microwave electromagnetic fields. Fiber-optical temperature probe is the most usual method for temperature measurement under electromagnetic fields. Also, domestic use of microwave ovens requires control and recommendations for consumers because the industrial use of microwave energy is typically monitored. Consumers need well-defined guidelines on the used technique for food preparation, in a way which minimizes the risk of formation of excessive harmful compounds.

References

Aktas, N., Ozligen, M., 1992. Injury of *E. coli* and degradation of riboflavin during pasteurization with microwaves in a tubular flow reactor. Lebensmittel-Wissenschaft und -Technologie 25, 422–425.

Albi, T., Lanzon, A., Guinda, A., Leon, M., Perez-Camino, M., 1997. Microwave and conventional heating effects on thermo-oxidative degradation of edible fats. Journal of the Science of Food and Agriculture 45, 3795–3798.

Ancos, B., Cano, M., Hernandez, A., Monreal, M., 1999. Effects of microwave heating on pigment composition and color of fruit purees. Journal of Foods Science and Agriculture 79, 663–670.

Barcia, M.T., Pertuzatti, P.B., Rodrigues, D., Bochi, V.C., Hermosin-Gutierrez, I., Godoy, H.T., 2015. Effect of drying methods on the phenolic content and antioxidant capacity of Brazilian winemaking byproducts and their stability over storage. International Journal of Food Sciences & Nutrition 66 (8), 895–903.

Bookwalter, G., Shulka, T., Kwolek, W., 1982. Microwave processing to destroy Salmonellae in corn-soy-milk blends and effect on product quality. Journal of Food Science 47, 1683–1686.

Bowers, J.A., 1972. Eating quality, sulfhydryl content, and TBA [2-thiobarbituric acid] values of Turkey breast muscle. Journal of Agricultural and Food Chemistry 20 (3), 706–708. https://doi.org/10.1021/jf60181a014.

Brewer, M., Begum, S., 2003. Effect of microwave power level and time on ascorbic acid content, peroxidase activity and color of selected vegetables. Journal of Food Processing and Preservation 27, 411–426.

Brown, E.F., Gonzalez, R.R., Burkman, T., Perez, T., Singh, I., Kristin, J., Reimers, K., Birla, S.L., 2020. Comparing nutritional levels in a commercially-available single-serve meal using microwave versus conventional oven heating. Journal of Microwave Power & Electromagnetic Energy 54 (2), 99–109. https://doi.org/10.1080/08327823.2020.1755483.

Canas, S., Perez-Moral, N., Edwards, C.H., 2020. Effect of cooking, 24 h cold storage, microwave reheating, and particle size on in vitro starch digestibility of dry and fresh pasta. Food & Function 11, 6265.

Clare, D., Bang, W., Cartwright, M., Drake, M., Coronel, P., Simunovic, J., 2005. Comparison of sensory, microbiological, and biochemical parameters of microwave versus indirect UHT fluid skim milk during storage. Journal of Dairy Science 88, 4172–4182.

Cremer, M., 1982. Sensory quality and energy use for scrambled eggs and beef patties heated in institutional microwave and convection oven. Journal of Food Science 47, 871–874.

Cross, G.A., Fung, D.Y.C., 1982. The effect of microwaves on nutrient value of foods. CRC Critical Reviews in Food Science & Nutrition 16, 355–382.

Datta, A.K., Hu, W., 1992. Optimization of quality in microwave heating. Food Technology 46 (12), 53–56.

Datta, A.K., Sahin, S., Sumnu, G., Keskin, S.O, 2007. Porous media characterization of breads baked using novel heating modes. Journal of Food Engineering 79 (1), 106–116.

Decareau, R.V., 1994. The microwave sterilization process. Microwave World 15 (2), 12–15.

Farag, R.S., Hewedp, F.M., Abu-Raiia, S.H., El-Baroty, G.S., 1992. Comparative study on the deterioration of oils by microwave and conventional heating. Journal of Food Protrction 55, 722–727.

Guan, D., Gray, P., Kang, D.H., et al., 2003. Microbiological validation of microwave-circulated water combination heating technology by inoculated pack studies. Journal of Food Science 68, 1428–1432.

Hafez, Y., Mohamed, A., Hewedy, F., Singh, G., 1985. Effects of microwave heating on solubility, digestibility and metabolism of soy protein. Journal of Food Science 50, 415–417.

Henderson, H.M., Hergenroeder, K., Stuchly, S., 1975. Effect of 245 MHz microwave radiation on horseradish peroxidase. Journal of Microwave Power 10 (1), 27–35.

Harrison, D., 1980. Microwave versus conventional cooking methods: effects on food quality attributes. Journal of Food Protection 43 (8), 633–637.

Hu, et al., 2021. Microwave technology: a novel approach to the transformation of natural metabolites. Chinese Medicine 16, 87. https://doi.org/10.1186/s13020-021-00500-8.

Jiang, H., Liu, Z., Wang, S., 2017. Microwave processing: effects and impacts on food components. Critical Reviews in Food Science and Nutrition. https://doi.org/10.1080/10408398.2017.1319322.

Kamat, G., Laskey, J., 1970. Enzyme Inactivation in Vitro with 2450 MHz Microwaves. Radiation Bio-Effects Summary Reports. US Department of Health. Education and Welfare, Publ. No. BRH/DBE 70-7.

Kermasha, S., Bisakowski, B., Ramaswamy, H.S., Van de Voort, F.R., 1993a. Comparison of microwave, conventional and combination treatments inactivation on wheat germ lipase activity. International Journal of Food Science and Technology 28, 617–623.

Kermasha, S., Bisakowski, B., Ramaswamy, H.S., Van de Voort, F.R., 1993b. Thermal and microwave inactivation of soybean lipoxygenase. Lebensmittel-Wissenschaft und -Technologie 26, 215–219.

Koutchma, T., Schmalts, M., 2002. Degradation of vitamin C after alternative treatments of juices. In: Institute of Food Technologies (IFT), 2002 Annual Meeting. Anaheim, CA. June 16–19.

Lau, M.H., Tang, J., 2002. Pasteurization of pickled asparagus using 915 MHz microwaves. Journal of Food Engineering 51, 283–290.

Li, N., Wang, L., Zhao, S., Qiao, D., Jia, C., Niu, M., Zhang, B., 2020. Multiscale structural disorganization of indica rice starch under microwave treatment with high water contents. Food Hydrocolloids 103, 105690.

Mapengo, C.R., Emmambux, M.N., 2022. Processing technologies for developing low GI foods—a review. Starch 2022, 2100243. https://doi.org/10.1002/star.202100243.

Massoubre, E., 2003. Foie gras et plats cuisinés: micro onde pour une bonne qualité sensorielle. Viandes et Produits Carnes 23, 49–52.

Mayeaux, M., Xu, Z., King, J., Prinyawiwatkul, W., 2006. Effects of cooking conditions on the lycopene content in tomatoes. Journal of Food Science 71 (8), C461–C464.

Michalak, J., Czarnowska-Kujawska, M., Klepacka, J., Gujska, E., 2020. Effect of microwave heating on the acrylamide formation in foods. Molecules 25, 4140. https://doi.org/10.3390/molecules25184140.

Murcia, M.A., Martinez-Tome, M., del Cerro, I., Sotillo, F., Ramirez, A., 1999. Proximate composition and vitamin E levels in egg yolk: losses by cooking in a microwave oven. Journal of the Science of Food and Agriculture 79, 1550–1556.

Nezihe, A., Elif, D., Ozlem, Y., Tuncer, E.A., 2011. Microwave heating application to produce dehydrated castor oil. Industrial & Engineering Chemistry Research 50, 398–403.

Okmen, Z., Bayindirli, A.L., 1999. Effect of microwave processing on water soluble vitamins: kinetic parameters. International Journal of Food Properties 2 (3), 255–264.

Ohlsson, T., 1987. Sterilization of foods by microwaves. In: International Seminar on New Trends in Aseptic Processing and Packaging of Foods Stuffs. Munich, October 2223.

Ovesen, L., Jakobsen, T., Leth, T., Reinholdt, J., 1996. The effect of microwave heating on vitamins B1 and E, and linoleic and linolenic acids and immunoglobulins in human milk. International Journal of Food Sciences & Nutrition 47 (5), 427–436.

Ramaswamy, H., Fakhouri, M., 1998. Microwave blanching: effects on peroxidase activity, texture and quality of frozen vegetables. Journal of Food Science & Technology 35 (3), 216–222.

Rodrieguez-Lopez, J., Fenoll, L., Tudela, J., Devece, C., Sanchez-Hernandez, D., Reyes, E., Garcia-Canovas, F., 1999. Thermal inactivation of mushroom polyphenoloxidase employing 2450 MHz microwave radiation. Journal of Agricultural and Food Chemistry Chem47, 3028–3035.

Rydberg, P., Eriksson, S., Tareke, E., Karlsson, P., Ehrenberg, L., Törnqvist, M., 2005. Factors that influence the acrylamide content of heated foods. In: Chemistry and Safety of Acrylamide in Food; Springer: Boston, MA, USA, pp. 317–328.

Szepes, A., Hasznos-Nezdei, M., Kovacs, J., Funke, Z., Ulrich, J., Szaboreesz, P., 2005. Microwave processing of natural biopolymers-studies on the properties of different starches. International Journal of Pharmacy 302, 166–171.

Tajchakavit, S., Ramaswamy, H., 1995. Continuous-flow microwave heating of orange juice: evidence of non-thermal effects. Journal of Microwave Power & Electromagnetic Energy 30 (3), 141–148.

Tajchakavit, S., Ramaswamy, H.S., 1996. Thermal vs. microwave inactivation kinetics of pectin methylesterase in orange juice under batch mode heating conditions. Lebensmittel-Wissenschaft und -Technologie 2, 85–93.

Tajchakavit, S., Ramaswamy, H.S., 1997. Continuous-flow microwave inactivation kinetics of pectin methylesterase in orange juice. Journal of Food Processing and Preservation 21, 365–378.

Tang, J.J., 2015. Unlocking Potentials of Microwaves for Food Safety and Quality Journal of Food Science Special Issue: 75 Years of Advancing Food Science, and Preparing for the Next 75.. 80.. 2015. https://doi.org/10.1111/1750-3841.12959.

Tong, Z., Koutchma, T., Ramaswamy, H.S., 2000. Evaluation of the relative effectiveness of continuous-flow microwave heating using time-temperature integrators: 3 p. In: 35th Annual Symposium of Institute of Microwave Power, July 17-19, 2000, Montréal, Québec, Canada.

Turkmen, N., Sari, F.Y., Velioglu, S., 2004. The effect of cooking methods on total phenolics and antioxidant activity of selected green vegetables. Food Chemistry 93 (4), 713–718. https://doi.org/10.1016/j.foodchem.2004.12.038.

Vikram, V.B., Ramesh, M.N., Prapulla, S.G., 2005. Thermal degradation kinetics of nutrients in orange juice heated by electromagnetic and conventional methods. Journal of Food Engineering 69, 31–40.

Wang, S.H., Toledo, M.C., 1987. Inactivation of soybean lipoxegenase by microwave heating: effect of moisture content and exposure time. Journal of Food Science 52 (5), 1344–1347.

Watanabe, F., Abe, K., Fujita, T., Goto, M., Hiemori, M., Nakano, Y., 1998. Effects of microwave heating on the loss of Vitamin B12 in Foods. Journal of Agricultural and Food Chemistry 46, 206–210.

Welt, B., Tong, C., 1993. Effect of microwave radiation on thiamin degradation kinetics. Journal of Microwave Power & Electromagnetic Energy 28 (4), 187–195.

Yaylayan, V., Roberts, D., 2001. Generation and release of food aroma. In: Datta, A. (Ed.), Handbook of Microwave Technology for Food Applications. Marcel Dekker, Inc., pp. 173–186

Yoshida, H., Takagi, S., 1996. Vitamin E and oxidative stability of soya bean oil prepared with beans at various moisture contents roasted in a microwave oven. Journal of the Science of Food and Agriculture 72, 111–119.

Yuan, Y., Zhao, G., Chen, F., Liu, J., Zhang, H., Hu, X., 2007. A comparative study of acrylamide formation induced by microwave and conventional heating methods. Journal of Food Science 72 (4), C212–C216.

Zeng, S., Chen, B., Zeng, H., Guo, Z., Lu, X., Zhang, Y., Zheng, B.J., 2016. Structural characteristics and physicochemical properties of lotus seed resistant starch prepared by different methods. Food Chemistry 64, 2442.

CHAPTER 5

Essential aspects of commercialization of applications of microwave and radio frequency heating for foods

5.1 Introduction

Microwave processing is a well-accepted technology for domestic applications and certain industrial operations for the past 70 years. The use of microwave heating systems demonstrated its advantages for the treatment of some products when other products require more research in order to accelerate commercialization. Its use in many food processing steps, such as cooking and blanching, drying, tempering, and preservation, is marketed and well projected for the future. As an example, continuous microwave cooking systems are used to cook meats commercially. Bacon precooking and crisping is one of the most common operations of industrial microwave processing. Many fast-food restaurants are commercial users of this technology. Microwave cooking is also used as a booster prior to an impingement oven for fully cooked, bone-in chicken. Conventional cooking can leave blood spots or cause overcooking on the exterior of meat, whereas microwaves penetrate to start internal cooking more quickly, reducing cooking time by almost 50% and increasing throughput by 30%. Boneless meatball cooking is another application of microwave boosters. Also, microwave drying is very efficient and is commercially used in applications for cookies, snacks, as well as spices and other ingredients. It is also used to finish drying process of pasta and instant noodles. It is estimated that microwave use as a predryer or postdryer can increase overall production capacities by 25%–33% and can produce a return on capital investment within as little as 12–24 months. The common application of microwaves in food processing is for tempering of meat, poultry, seafood, fish, or frozen fruits or vegetables. Microwaves tempering can be achieved in minutes rather than hours required using air or water-based systems even for large product blocks. Tempering can also be performed directly inside the package. This process results in a significant reduction in drip loss while minimizing product deterioration due to bacterial growth, making it ideal for many defrosting processes.

Industrial microwave heating systems for the purpose of food pasteurization or sterilization are not yet fully commercialized except for a few product categories. The lack of uniformity of heating of solid prepackaged foods and methods to control it has been a significant technical hurdle. Uniformity is easier to achieve using continuous flow microwave heating process of liquid and semiliquid products through product mixing in turbulent regime and applying principles of aseptic packaging. Reliable methods for the temperature measurements and quantification of heating uniformity are a prerequisite in order to achieve the success of microwave processing concepts with regard to a better heating uniformity and process control. The design of packaging and materials is another essential component in microwave treatment of in-package products. Because of its benefits, the microwave process has a significant potential as a preservation technology for various foods and ingredients and may be successfully used in combination with conventional heating modes and other physical methods. This chapter will discuss the aspects of microwave processing technology that are essential for its successful commercialization such as packaging, temperature measurement, validation approaches in commercial systems and domestic ovens, and use of computer simulation to improve microwave process heating performance.

5.2 Packaging for microwave heating

Microwave package can be used for cooking or heating the product in microwave systems. Packaging materials react to microwaves in three ways: transmit, reflect, and absorb the radiation. By proper material selection and package design, the amount of energy transmitted, reflected, and absorbed can be varied, controlled, and proportioned. In addition, thermal properties of the package affect the heat transfer between food and microwaves and can modify the heating pattern of the food by releasing the water vapors inside the package.

Based on material properties, there are two types of microwave packaging materials that allow microwaves to pass through the material called microwave transparent passive packaging and microwave active packaging that allows using of packaging material and directly affect the cooking of the product inside the container.

5.2.1 Passive packaging

Most of the conventional packaging materials are transparent to microwave, and the product can be heated without interfering with these packaging materials including:

- Glass and ceramics
- Paper, paper board, and structured films
- Polymer materials and trays
    - polyethylene (PE), polypropylene (PP), polyester, nylon, polystyrene, and poly vinyl chloride
- Edible materials

The selection of appropriate packaging materials depends on many factors, such as type of food, shelf life and storage conditions, product positioning as value or premium, in-home or out-home, on-the-go or restaurant. Their application in microwave package development depends upon their thermal stability and compatibility with the product under high temperature and pressure, and packaging should perform all the functions of a conventional package. The essential packaging requirements for microwave heating are as follows:

- Allow fast microwave heating
- Barrier to oxygen, moisture, and microorganisms
- Compatibility with product
- Withstand processing and storage conditions
- Retain solvent and odor in packaging material during storage and cooking
- Provide physical strength during processing, transportation, and storage
- Cost-effective
- Uniform distribution of heat within the pack
- Safety of the consumer
- Insulated label
- A means for holding the pack after reheating

Glass and glass jars or containers are suitable for microwave heating that depends on shape, size, and volume of the container, initial temperature, and properties of food. It was demonstrated to a salsa manufacturer that they can use his existing glass jar with a metal cap. The metal cap actually proved to be an advantage. By adjusting the position of the magnetrons, the salsa was cooked from the bottom of the jar, with the beams reflecting back from the cap, resulting in a faster cooking process.

Among numerous plastic materials, high-density polypropylene (HDPP) is a low-cost solution for pasteurization, as are other materials that can withstand the target temperature. For sterilization, polyethylene terephthalate (PET), HDPP, and various polyester-based materials are available as high-quality trays, pouches, and bags.

Volatile compounds can give off-odor and off-taste to foodstuffs when heated, even if the material had been found to be suitable for extended storage at room temperature. Migration tests with aqueous simulants in a heating chamber were conducted with plastics used for packaging. The materials had considerable differences with regard to the content of volatile compounds. A PET coating prevented the migration of the off-taste, whereas PE and PP were only partially effective. With these two plastics, the thickness of the plastic layers also affects the amount and level of off-taste. The highest off-taste levels in foodstuffs were found on PP and polyvinylidene chloride plastic. The plastic coating can serve as a barrier to the transfer of off-taste into foodstuffs.

5.2.2 Active packaging

Active food packaging can affect and control the cooking process during microwave heating. The rationality is that the heating is not only related to the power source but also to the food itself, including properties and geometry, which are factors of how much microwave energy is transferred to foods. In active packages, the microwaves are modified by the use of susceptors, reflectors, or a guidance system. As an example, a round container may need different heating time and energy compared to rectangular container with the same amount of same food product because container itself adds to the effectiveness of microwave heating process. The size shape, location, and composition of microwave active package can alter the environment both inside the package and in the microwave cavity. Also, the food that are unsuitable for reheating in the microwave oven use these packaging materials. The active packaging can be separated on the regular packaging and packaging that assist the heating either by providing heated surfaces (susceptors) or by modifying the fields through metal shielding acting as field modifier from the packaging. Therefore, three types of active packaging are available.

- Metalized paper (susceptor) or other materials that can create heated surfaces in the process of heating.
- Metallic materials that define the electromagnetic boundary conditions in the cavity.
- Dielectric materials or components that can affect the heating patterns in the product.
- Regular containers that only define the geometry of the food product.

Aluminum trays are suitable for heating both frozen and refrigerated foods in a microwave oven when they are correctly used. Aluminum trays are also suited for heating in combined microwave and convection ovens, whereas plastic dishes can only be used for certain foods. The postheating quality of casserole and pizza can be improved by packaging them in an aluminum tray instead of a plastic tray. The aluminum trays should be positioned so that they do not touch the walls in the oven. However, aluminum foils are used to cover the food sparked even when they did not touch the walls. Packages containing susceptors didn't catch fire even during prolonged heating in a microwave oven. However, the food is charred and the walls of the oven are covered with tar.

5.2.3 Microwavable packages

A few types of the microwavable packages are currently used in the food industry.

5.2.3.1 Flexible microwavable packages and pouches

Microwavable flexible packages in the form of bags, overwraps, sleeves, and boil-in-bag are made from susceptors, laminated to paper and other high heat-resistant plastic materials. In other words, packaging materials are incorporated with microwave-active components in roll stock form.

5.2.3.2 Microwave safe coated aluminum tray

Aluminum trays are coated with vinyl and epoxy resins to absorb microwave energy and prevent arcing. They are sealed with transparent lidding material or foil laminates. A snap on reclosable plastic dome is used for microwave heating. They can withstand the freezer temperature to microwave or conventional oven heating without losing firmness.

5.2.3.3 Rigid plastic tray
Rigid plastic trays are made from the plastic materials, which can be used in combination or as monolayers, such as PET, PP, nylon, polycarbonate (PC), polysulfone, high-density polyethylene, etc. They are sealed by heat-sealable lidding materials, overwrapped/shrink wrapped, or sealed inside a microwavable bag. Because of their inline formability into different attractive shapes and sizes, easy availability of material, and possibility of blending different materials of different performance and cost parameters to meet the end use make them more attractive.

5.2.3.4 Paper board containers
Polyester-coated paperboard containers are most common microwavable packages that can be molded in different shapes and sizes. They are produced on a conventional tray making, carton forming, and folding carton making machine and are the least expensive microwave transparent material.

5.2.3.5 Crystallized polyester
Crystallized polyester containers are also popular for applications as microwave packaging as well as for conventional cooking. Crystallized polyester is vacuum formed from shed stock into different shapes, is stiff, and has a good appearance. The containers are sealed with transparent/nontransparent lidding material in a high-speed tray sealing machine, are easy to handle, sturdy, attractive, cost-efficient, and can be compartmentalized for multicomponent food.

5.2.3.6 Molded pulp containers
Molded pulp trays are stronger than pressed trays. They are very stiff and have quality appearance. These containers are economical and are made in attractive designs and colors. Due to their strength, appearance, and compatibility, they are used as self-serving containers.

5.2.3.7 Nonself-venting and self-venting materials
The heating of microwavable products can be enhanced by secondary heating effect, such as the steam generated within the package or bag by microwave energy from water or high-water content foods. Steam is characterized by effective heat transfer properties and can be used to achieve more desirable food attributes. At home, water can be added in a container of multiple compartments or have more water in foods. Since water has high dielectric loss factor, it will absorb microwave energy and create steam first.

Nowadays, the increasing demand for ready-to-cook or ready-to-eat (RTE) products led to optimizing typical packaging solutions to create systems that can provide heating and release steam during microwaving. New functional packaging materials for microwave applications can be divided into two categories: nonself-venting and self-venting materials.

Nonself-venting materials are materials that do not allow the release of internal steam during microwave heating. Therefore, consumers have to open up covers or puncture packages before microwaving product to facilitate the release of internal steam.

Self-venting materials are smart materials that can safely release the internal steam or pressure built up from inside the packages without necessitating opening prior to microwaving.

Microwave packaging materials and technologies incorporating self-ventilation features are studied and commercialized for microwave packaging application. Thailand's TPN FlexPak developed the "M-Vent Pouch" packaging. This packaging presents a novel sealing feature that allows it to "breathe out" while food is microwaved. It permits the safe release of steam from inside the package without requiring partial opening or puncturing before placing it in a microwave oven. This pouch was manufactured by combining PE and PP materials into a multilayer structure. This structure comprises a PE outer layer laminated to a PET substrate and a Borclear PP carrier layer, which delivers good heat resistance and robust sealing. Such a structure can be easily processed on existing PE blown film lines, and therefore, does not require investing in new production equipment.

Recently, controllable permeability technologies, such as polymers exhibiting temperature-dependent permeability have been applied in many fields, such as membranes, drug delivery systems, and packaging. Knowledge of the temperature dependence of gas or water vapor permeability of polymer materials is crucial for food packaging exposed to different or uncontrollable environmental conditions.

5.3 Microwave process validation

Validation is a preemptive scientific evaluation that provides documentary evidence that a particular process (e.g., cooking, frying, chemical treatment, extrusion, etc.) is capable of consistently delivering a product that meets predetermined specifications. A successful validation study requires diverse expertise, detailed design, an experienced microbiologist, a statistician, and a process authority.

Because of the thermal nature of microwave heating, microbiological validation tests are conducted using procedures that are similar to those used for thermal processing. The main challenge of microwave heating is that microbial survival of various microwave treatments was associated to the nonuniformity of the temperature distribution during the process. The lack of uniformity of heating has been a significant technical hurdle. Achieving heating uniformity remains a major challenge in the research and development of microwave heating preservation technologies. The nonuniform heating can result from the varying absorbing dielectric properties of foods and the surrounding air as well as from the difference in the dielectric properties of different food constituents and their dependency on temperature and packaging. Due to these challenges, the approaches for microwave process validation can differ from traditional heating process and based on different ways to demonstrate and document targeted performance.

Characterization of the heating parameters of the microwave systems is the essential first step needed to determine efficient processing conditions and select optimal treatment. Heating characteristics of microwave processing system as it was discussed in chapter 2 include:

- spatial and time-temperature relationships within the product during transient and steady-state heating periods
- heating rates per unit of time
- absorbed power
- coupling efficiency defined as a ratio of absorbed energy and heat losses to nominal incident power.

Knowledge of spatial temperature distribution is critical for the determination of a location of the least heated point or the so-called "cold spot" in the product and further calculation of process lethality. In a case of continuous flow process, temperature rise versus flow rate/residence time and input power need to be known. In addition, flow regime (laminar or turbulent) and residence time distribution in microwave cavity need to be determined.

Also, it has been reported that there was no microorganism found with the unique resistance to microwave processing, suggesting that classical surrogates (vegetative cells or spores) would be appropriate for process determination and validation. A few surrogates have been recommended for commercially sterilized low-acid (pH > 4.6) foods. Putrefactive anaerobe *Clostridium sporogenes* strain PA 3679 has been used as a nontoxigenic surrogate for proteolytic *Clostridium botulinum* in the validation of conventional heat sterilization processes. It too would be used for the derivation of thermal processes for microwave sterilization systems as a surrogate to produce a safe product (Guan et al., 2003; Tang, 2015). The reported D-value at 121°C of 0.6 min for *C. sporogenes* PA 3679 is higher than that of proteolytic *C. botulinum* (D at 121°C = 0.24 min), and both organisms have a z-value close to 10°C. Thus, thermal processes based on sufficient inactivation of spores of PA 3679 typically provide a considerable safety margin with respect to the destruction of spores of proteolytic *C. botulinum*.

Listeria monocytogenes has been commonly used as a target pathogen for designing microwave pasteurization processes (Tang, 2015) because of its greater thermal resistance compared to *Escherichia coli*, *Salmonella enteritidis*, or *Salmonella typhimurium*. Nonpathogenic *Listeria innocua* strains have been used as surrogates for *L. monocytogenes* because of their behavioral and resistance similarities.

Packaging validation is a part of microwave process validation. The microwave processing effects on sealing, physical structure, and oxygen permeation properties of packaging materials have to be determined along with a probability of migration of packaging chemical compounds into the product. Shelf life studies have to be conducted, and quality aspects of both product and packaging have to be estimated at storage temperature during planned storage period.

5.4 Temperature and process lethality measurements during microwave heating

The biggest challenge of microwave validation testing is measuring a temperature history in time and space and establishing cold spot within the system and the product. Temperature control systems are needed on a production scale, pilot and laboratory scale, where the process and product requirements are defined and have to be validated. Temperature measurement single point probes, radiometry and magnetic resonance imaging 2D measurement system, chemical or biological markers, and computer simulation that allow obtaining 3D temperature mapping and corresponding lethality distribution are current approaches to determine the temperature profiles and cold spots in microwave processing.

5.4.1 Temperature probes

Researchers also have identified fiber-optic sensors suitable for monitoring the rapid temperature rise that will occur in the microwave chamber. Fiber-optical sensors allow avoiding interference from microwave fields and often used for temperature spot measurement inside the sample covering a temperature range from −200 up to 450°C. Temperature measurement by fiber-optic sensors is based on fluorescence decay time, Fabry—Perot interferometry, or transmission spectrum shift in semiconductor crystals. The physical base of fiber-optic temperature measurement principle is the determination of the temperature-dependent Raman scattering in a glass fiber. As input signal into the glass fiber, a laser, a xenon flash lamp, or light-emitting diodes emit light pulses. At the tip of the flexible fiber, the components that are reactive to temperatures, such as phosphors or liquid crystals are incorporated. If light hits the excited molecules, photons and electron interact (photoluminescence), generating temperature-dependent light scattering or Raman scattering. The signal conditioner detects the changes of the back-scattered light and converts it into an output signal. Measurement accuracy of 0.1°C is reported for the temperature range of food processing applications. The limitations of fiber-optic probes are in their usage because the sensors are mechanically sensitive, and they limit the movement of samples. Accurate readings of product temperature can be achieved when the probe is in direct contact with the sample. Additionally, the exact positioning of the fiber probe can be complex, especially in porous or structurally sensitive materials. Steam bubbles created by the evaporation of water can disturb and misrepresent the temperature measurement. This affects the spatial resolution of the method.

Some fiber-optic sensors on the market can measure temperature along the entire length of the fiber with resolutions in the millimeter range by using the light scattered back along the fiber material itself and allow detecting more data points.

Temperature measurement during microwave heating is crucial due to possible nonuniform temperature distribution, a fiber-optic sensor might not fulfill the requirements for microwave application such as pasteurization or sterilizations. The issues regarding spatial resolution, correct positioning, and fixation of the probe's present specific difficulties in practical handling of fiber-optic sensors.

The possibility of using metallic mobile temperature sensors that are commercially available was explored for conventional continuous canning operations and reported by Tang and Liu (2015). Metal thermocouples can induce thermal instabilities, microwave breakdown, distortion of the microwave field, and result in measurement errors and sensor destruction. However, their application is possible if the sensor components are electrically shielded from the microwave field. Using aluminum tube and thermal insulation was an initial solution that required a direct connection to a signal conditioning and data acquisition unit outside the microwave cavity, limiting sample movement. This temperature measurement method does not work for continuous processes where food packages have to move in the system's chamber. This issue has been addressed by the application of wireless mobile temperature sensors. They act as stand-alone, fully functional temperature sensors including data logger and sensor probe that can be read out after the microwave processing and do not limit continuous sample movement.

5.4.2 Thermal imaging

Thermal imaging is a contactless, noninvasive, infrared (IR) technology that uses IR-sensitive cameras to measure the surface temperature distribution of the product without affecting the electromagnetic field (EMF) within the microwave treatment chamber. The spatial resolution depends on the optical density of the camera and the distance to the heated object. The IR thermography can be a useful tool for measuring the sample surface temperature in 2D and demonstrate the evolution of temperature change. This is a noncontact type of temperature transducer. However, the accuracy of IR sensors is limited and does not directly allow to measure the evaluation of cross-sectional temperature profiles.

Similar to IR cameras, pyrometers are based on the detection of IR radiation emitted by the heated products and only measure surface temperatures. This limits their application to some extent for the detection of hot and cold spots in the inner core of the material and automated temperature-controlled microwave processes based on pyrometer data. Therefore, pyrometers in microwave applications are mainly suitable for controlling homogeneous temperatures or temperatures at predefined critical control points where single spot measurement is not a limitation.

In thermal paper upon reaching a specific trigger temperature, a single irreversible color change can be induced and used to qualitatively test the uniformity of energy distribution in microwave cavities. Thermal paper can be used as a method to compare simulated and experimental results regarding energy distributions in microwave cavity or the temperature distribution at the surface of microwave-treated products.

5.4.3 Process lethality indicators

The basic research has focused considerable effort on developing predictive methods to quickly and reliably determine the location of cold spots. However, the process lethality is a function of temperature and time. This means that time/temperature-indicators (TTIs) are required for a complete characterization of microwave process and quantify process lethality (F-value). Two types of TTI chemical (Maillard reaction or sugar caramelization) and biological (microbiological and enzymatic) indicators have been developed and tested. Three Maillard reaction intermediates have been shown to be suitable as chemical markers to measure heating effects and heating uniformity in microwave processing of food (Kim and Taub 1993; Kim et al., 1996). These chemical markers were entitled as M-1 (2,3-dihydro-3,5-dihydroxy-6-methyl-4(H)-pyran-4-one), M-2 (4-hydroxy-5-methyl-3(2H)-furanone), or M-3 or HMF (5-hydroxymethylfurfural). The mechanism of action of these markers is based on color change due to the heating of model food system in which the physical and dielectric properties are similar to the properties of tested foods. The color of those model food systems is light white before thermal processing. When these models are heated in a package with microwaves, localized brown color has been developed. The color intensity depends upon the intensity of heating. The overall recommendation is that the M-1 marker is mainly appropriate for monitoring pasteurization processes (Tang et al., 2007), and the M-2 marker is suitable for microwave sterilization (Wang and others 2018). As an example, Tang (2015) reported a development and application of standard whey protein gels and mashed potatoes models used with chemical marker precursors M-2 that cover the range of dielectric and thermal properties similar to those of different types of foods. Direct correlations were established between local thermal lethality F-value and color intensity as a result of M-2 formation from microwave sterilization. The chemical marker M-2 was used with digital imaging and Image Acquisition Vision Builder software to generate 3D heating patterns of food. So far, there are no reports available regarding practical application of the markers in the analysis of microwave processes.

Examples of microbiological TTIs are spores from specific spore formers, such as *Bacillus atrophaeus or Geobacillus stearothermophilus*. Commercially available microbiological TTIs may differ broadly and even vary between batches of the same supplier in their thermal resistance; there is no established routine method for the production of microbiological TTIs.

Enzymatic TTIs have been used for time-temperature history and thermal performance monitoring. The principle of enzymatic TTIs is based on the irreversible change of the enzymes' structure and decreasing their activity due to heating. The enzyme activity loss can be measured to calculate the time-temperature history of the sample as a result of heat processing. Polyphenol oxidase and peroxidase enzymes have been tested as enzymatic TTIs because they are comparably heat-resistant. In addition, other types of enzymes, such as α-amylase and alkaline phosphatase, may also be used as enzymatic TTIs.

5.5 Microwavable foods and cooking instructions

Microwaveable foods are intended to be heated or reheated by the use of a microwave oven and developed to satisfy the demands of faster and convenient life styles of consumers when products can be prepared in a short time. The key advantage of using microwave oven over conventional heating oven is that typical product preparation in microwave appliances requires much less time. In the case of microwaveable foods, the product can be heated or cooked in shorter times while maintaining good quality characteristics such as texture, nutrients content, and flavor, very similar to products cooked conventionally. Currently, microwaveable products occupy a significant proportion (87.5%) of the frozen and refrigerated foods sections in supermarkets and grocery stores.

Microwaveable foods can be classified according to the time required for heating the product. Fully cooked and precooked RTE microwaveable products include foods that only need to be only reheated during a short time in a microwave oven before consumption. The examples of such microwavable meals include fully cooked snacks, precooked entrees, sandwiches, pizzas, or one serving size meals. Heating instructions for these products are developed and provided by the product manufacturer to assure that the quality and safety of the final product is similar to the one for conventionally cooked food. It is important to emphasize that thermal processing applied during manufacturing of these foods was targeted and implemented to eliminate foodborne pathogens such as *E. coli*, *L. monocytogenes*, and *Salmonella* spp. in the final product.

On the other hand, not-ready-to-eat (NRTE) foods are mixtures of meat or poultry with any other ingredient such as vegetables in which at least one ingredient has not received a heat treatment for the elimination of pathogenic bacteria (GMA, 2008). The examples of NRTE foods include products such as frozen-stuffed chicken, pies, and raw chicken nuggets. The USDA-FSI classified NRTE products as raw, and they are considered as a potential source for the transmission of pathogenic bacteria. The presence of raw ingredients, regardless of the component (meat, poultry, or vegetables), implicates that the consumer must fully cook the product through the exposure to longer heating time for the elimination of potential foodborne pathogens that may grow in the product. Therefore, appropriate product heating procedures in microwave oven should be applied to NRTE foods in order to reduce the risk of transmission of foodborne illness.

Following to the outbreaks for the consumption of NRTE frozen food products, the USDA-FSIS issued a letter to food processors that manufacture these types of products requiring the reevaluation and validation of the heating instructions for the lethality of *L. monocytogenes* or *Salmonella* spp. using microwave ovens.

To address this issue, the Grocery Manufacturer's Association (GMA) and the American Frozen Food Institute (AFFI) developed guidelines for food processors on the proper labeling of heating instructions and their validation (GMA, 2008). The AFFI developed guidelines for food processors on developing heating instructions for microwaveable food products (AFFI, 2008). These guidelines are recommendations and are not specific for all the products. However, food processors have to develop instructions for specific products considering intrinsic characteristics (volume, size, and composition). The GMA (GMA, 2008) developed and issued guidelines that would help to develop microwave oven heating instructions that can contribute to the elimination of pathogenic bacteria if present in the product. The guidelines cover the key factors that should be considered by the product manufacturer when developing and validating microwave heating instructions for NRTE foods.

As was discussed in previous chapters, the kinetic parameters of thermal destruction of food pathogens based on the use of D-values (decimal reduction time) and z-values (pathogen thermal resistance) have been used for the evaluation of lethality of conventional heating. However, because of the nature of microwave heating, this approach is not fully applicable for microwaveable NRTE food products. According to the Lethality Requirements guidelines for RTE meat and poultry products (USDA-FSIS, 1999), for fully cooked products, the target internal temperature of 160°F should provide sufficient lethality to assure the safety of product cooked by consumers. For products containing poultry that is not fully cooked, an internal temperature of 165°F is required (GMA, 2008).

The GMA recommended that the lethality requirements may be different for NRTE products.

It is recommended that microbial challenge studies should be performed to validate the microwave heating instructions. The objective of the instruction's validation is to test and determine microwave heating times and temperatures that are adequate to eliminate pathogenic bacteria, intentionally introduced, in microwaveable, but NRTE products (GMA, 2008). For Salmonella spp., a 7 log CFU/g reduction is required to assure the food safety of RTE poultry products (USDA-FSIS, 1999). In order to determine the efficacy of the microwave heating instructions, the guidelines recommended the use of a minimum number of samples to evaluate the variability and performance of microwave heating. As the evaluation of the adequacy of time-temperatures regimes is performed, the food manufacturer has to identify the factors that affect microwave heating and contribute to the variability of the temperature of the product. Additional factors for evaluation that may have impact microwave heating are food packaging configuration and material or product type.

The following recommendations should be followed in order to properly conduct microwave oven validation tests.

1. The validation test should be performed using calibrated devices, appliances, and worst-case samples with sufficient replicates to allow an acceptable degree of confidence in the results. Any devices used to measure the temperature of a thermal process need to be calibrated on a regular basis against a certified reference thermometer. Calibration of devices should be performed depending on the initial temperature of the samples requiring testing, e.g., at frozen (−18°C), refrigerated (4°C), and at approximately 70°C temperatures. To ensure that probes are working correctly, it is helpful to perform an ice point/boiling point check (using iced water at 0°C and boiling water at 100°C) prior the tests.
2. The measurement of the power output of microwave oven also has to be done in the process of instructions validation. The standardized method of measuring

the power output is described in a section of the IEC 60705:2015 document. Simpler shortened methods of measuring the power output, e.g., using two 500 mL or one 1000 mL beakers of water can't be relied upon to give an accurate measurement result and should not be used. Indeed, calibrating ovens using a nonstandard method can give an oven power output that may be lower (or higher) than the actual power output. The power output of microwave ovens can be affected by the power supply voltage and the temperature of the microwave magnetron. Because the magnetron heats as the oven is used, the power output of the oven can decrease as the magnetron heats up. It can be suggested that microwave ovens should be preheated before instruction validation trials are performed. This would allow testing the instructions in the worst-case scenario where the microwave had been used previously and the power output had reduced.

3. It is essential that the variation in microwave oven heating performance is considered when selecting ovens for use in instruction validation. Microwave ovens of various designs and power outputs should be used for instruction validation trials. This will help to confirm that these ovens produce different microwave fields and temperature patterns inside the oven cavity and food product. Therefore, it will help to ensure that the validated heating instructions will be applicable to the wide range of domestic ovens. Other essential features that should be considered when selecting microwave ovens include: turntable or nonturntable, large or small cavity size, painted steel or stainless steel interior, heating modes combination (e.g., grill or hot air oven and microwave) or microwave only.

4. It is recommended that instruction validation should be performed to ensure the instructions will allow the delivery of the required minimum thermal process (or temperature) using worst-case product samples. Worst-case product samples are those defined as taking longest time to heat. Product samples chosen for testing should have: coldest initial temperatures likely to be found in consumers fridges or freezers 0–3°C for refrigerated products, −20 to −18°C for frozen foods; thickest product samples are likely to be found in the samples supplied to consumers; the test samples with the largest weight are likely to be found in the samples supplied to consumers.

5. Product initial temperatures must be stable throughout the tests, and this can be achieved by leaving samples to stabilize at the correct temperature throughout overnight. To ensure that the validated instructions will provide acceptable results for the broad range of production products, it may be helpful to also test slightly warmer, thinner, and lighter products from within the intended product supply range. Any changes to a product, recipe, size, shape, and packaging type would necessitate further instruction validation testing.

6. All information related to instruction validation trials should be documented in the report and retained for at least the life of the product.

5.6 Regulatory status and commercialization

In 2011, the United States Food and Drug Administration (US FDA) approved Microwave-Assisted Thermal Sterilization Process (MATS) using 915 MHz. As reported, the technology uses the immersion principle of packaged food in pressurized hot water while simultaneously heating it with microwaves at a frequency of 915 MHz. This combination allows to achieve higher temperature uniformity and eliminates food pathogens and spoilage microorganisms in just 5—8 min and produces safe foods with much higher quality than conventionally processed RTEs.

Wornick Foods, a manufacturer of convenience foods and customized meal solutions, was awarded a grant to establish MATS Research and Development Center at its facility in Ohio where consumer products companies can test the MATS process.

Microwave continuous flow sterilization of homogeneous and particulate-containing high and low acid viscous products has been developed using frequency of 915 MHz. The US FDA acceptance has been also granted in 2009 and cleared ways for commercial applications for pumpable low acid foods.

Radio Frequency Heating is the USDA Organic Thermal Process, US FDA Clean Label process, Food Safety Modernization Act (FSMA) compatible as a Critical Control Point Kill Step, and is able to be validated to comply with FSMA regulations.

Currently, there are no specific regulations on microwave food packaging. However, in the United States of America, any package that comes in contact with food must be suitable for the intended application under the US FDA good manufacturing practices regulations stipulated in Title 21 of the Code of Federal Regulations (CFR), Section 174.5 ("General provisions applicable to indirect food additives") (21 CFR 174.5. 174.5 1977; Packaging Law 2009). Even though no specific regulations on food packaging for microwave use have been enforced, the FDA had issued a guidance document "Guidance for Industry, Preparation of Premarket Submissions for Food Contact Substances: Chemistry Recommendations," which is intended for industry and contains the US FDA recommendations pertaining to the chemistry information that should be submitted in food contact notifications or food additive petitions for food contact substances (US FDA Guidance Documents, 2007).

The purpose of this guidance document is to explain the migration testing needed to clear a food contact substance for a particular intended use (US FDA Guidance Documents, 2007).

For microwave-only containers, the container is recommended to be tested based on the protocol for Condition of Use H (US FDA Guidance Documents, 2007). However, if the container is used as food contact article in microwave ovens, the migration test should be performed using food oils or fatty food simulants at 130°C (266°F) for 15 min and using aqueous food simulants at 100°C (212°F) for 15 min.

To ensure that microwave food packaging materials meet the relevant safety regulations, manufacturers should consider how the products are going to be used and which parts of the packaging will be in contact with the food product. Manufacturers should properly test packaging materials using relevant standard methods.

5.7 Industrial microwave processes and systems

Several companies manufacture microwave processing equipment for the food industry. A few of the leaders located in the United States are AMTek Microwaves, Cedar Rapids, Iowa, Industrial Microwave Systems (IMS), Morrisville, NC., Thermex-Thermatron, Louisville, KY., and Ferrite Microwave Technologies, Nashua, NH.

Multimagnetron tunnel ovens were widely used for pasteurization and sterilization. It is a cascade of many sections (the modular approach). In each section, IR, hot air, and water may also be applied for specific purposes. The transport system can be a conveyor belt, a quartz tube with configurations to improve heating uniformity. A microwave power consists of a magnetron, a wave guide section, radiation apertures, or an antenna. The magnetrons are controlled via PCs. When the material proceeds through the tunnel applicator, a power density is deposited internally and a corresponding temperature profile is a result. Any temperature profiles can be generated in almost all sorts of materials. Computer optimization methods exist to obtain the uniform temperature distribution in the product in the end of the tunnel. Temperature above 100°C needs creation of additional air pressure.

5.7.1 Ready-to-eat meals and in-pouch sterilization

Officine Meccaniche Attrezzature per Ceramiche (OMAC) (Avon, Mass.) and Berstorff Corp. (Charlotte, N.C.) were known as the premier suppliers of microwave processing systems for food in the world in the 1990s. Harlfinger L (1992) (Product Manager of OMAC Microwave Division, TW Kutter/Alfa Laval, Anon MA) reviewed the benefits of microwave sterilization and indicated that it can deliver good taste products because microwaves are able to heat the product three to five times faster than conventional systems. Microwave heating process was largely used for pasta, bakery, and prepared meals. The heating phase of the microwave sterilization cycle occurs in about 8–12 min, depending on the product characteristics, starting temperature, package weight, and dimensions. Equilibration, holding, and cooling will remain the same regardless of the sterilization method chosen. Equilibration will take about 2–3 min. Holding time depends on the F_o value required by the producer and will be usually 5–8 min. Cooling time will be equal to or longer that the heating time. A complete microwave sterilization system (TW Kutter's Pmac system) consists of a collating and infeed machine, microwave chamber, and a programmable logic controller. It is possible to use a microwave preheater between packaging and sterilization to even out temperatures or increase production from the sterilization unit. If the temperature difference ($T_f - T_o$) is lower, the energy requirement is also lower. The microwave sterilizer performed six functions: compression (2.5 bar), heating, equilibrating, holding (approximately at 260°F), cooling (using refrigerated circulated air down to 120–150°F), and decompression. The heating section used 1.9 kW magnetrons at 2450 MHz. The magnetrons surrounded the transport system, so the packages got microwave energy from the top and bottom of the conveyor. Each magnetron had variable power and controlled individually. The parameters of temperature, pressure, speed and cycle time, and electromagnetic power can be controlled. The technology has an ability to check that a certain F_o value has been achieved. Data trace recording device (Wheat Ridge, CO) can be installed inside the package, and it monitors temperature up to four points. Gourmet Fresh Pasta (GFP, Los Angeles) was using semirigid trays and getting 45-day shelf life for its fresh pasta product but had 5% returns due to spoilage. GFP converted from surface pasteurization and gas flushing to an OMAC microwave pasteurization system. The result was achieving 120-day shelf life marked for 90 days, with returns eliminated. OMAC claimed that return on investment can be two to three years.

Microwave processing for fresh filled pasta became common in Italy in the 1990s, and the technology had been applied to RTE ESL meals, pasta-based products, and a variety of other foods throughout Europe, Japan, and parts of South America. Some of the biggest processors in the world used the technology, including Unilever and Barilla SpA in Italy and Morinaga in Japan. The leading supplier of those systems was OMAC, established by engineer Giuseppe Ruozi to fabricate the multistage system that he designed. Systems were capable of processing up to 4400 lbs. of food per hour, but by 1995, OMAC was defunct. In 2000, Classica Microwave Technologies Inc. (NJ, USA) acquired OMAC's assets. The firm also retained the services of Ruozi as its chief technology officer and hired Joseph Riemer to serve as the president and execute the business plan. In 2003, Classica had almost 200 industrial installations in Europe but only two in the United States. Classica invested in a state-of-the-art lab in Sayreville, NJ., with pilot-plant equipment for microwave pasteurization, sterilization, sanitizing, and drying.

Classica developed a truly continuous process for pasteurization, with packaged product transported on a belt conveyor at a uniform, constant speed. After the product entered the tunnel, it reached the active microwave zone, where rapid heating occurred in synergy with hot air. The travel distance has been designed, so that the product attained the uniform, target temperature in the geometric space. Next, the product entered a hot-air-only zone, where temperature uniformity is guaranteed through the volume of the package. After a proper holding period, the product exited the chamber for water or air cooling.

A much more sophisticated system was designed for sterilization process with elevated pressure to be applied to prevent the package from being deformed. A continuous sterilization process was developed using a sequential multibatch approach with parallel lines. Product in the first line was subjected to rapid heating while the next line was loaded. Hydraulic gates seal the tubular tunnel, and counterpressure builds, while microwave heat and hot air brought the product up to the target temperature. By the time the last line has been loaded, the first one has completed the cycle and was ready to be reloaded. Product was rotated on its axis within the chambers, reducing processing time and guaranteeing temperature equilibrium. Food components within the package exchange heat until they reached the uniform temperature. After equilibration and holding time, the product moved under pressure to a cooling section. All these subprocesses took place in 7–15 min from the time the product entered the chamber. This system used magnetrons of 2 and 3 KW power to heat in-container food in high-efficiency cavities. The single system was constructed with 300 individual magnetrons.

Classica demonstrated to major bakeries that they could remove the preservatives in their formulas, pasteurize bread after it has been sliced and packaged, and create a product that is mold-free for more than 90 days at ambient temperature. With preservatives, mold began to grow within 21 days. The process couldn't stop bread staling; the bakers had to address the squeezability function. However, microwave pasteurization lowered returns for bakeries and, more importantly, gave them an opportunity to introduce high-end, value-added bread products with a "no preservatives added" statement on the package.

IMS acquired the assets of the now-defunct Classica Microwave Technologies Inc. in 2007. Besides a continuous system for pumpable foods, the company built ready a microwave unit for extended shelf life trials.

Berstorff Corp. (Charlotte, NC) built a pasteurization tunnel with 6–24 magnetrons (Schlegel, 1992). Each magnetron was controlled by a computer; the power was programmable between 10% and 100%. The system has been used to pasteurize or sterilize food products in trays. The achieved benefits were in improvement of quality, extension of shelf life without preservatives, lower distribution cost, energy savings, low maintenance cost, and minimal personnel. In addition, microwave technology is environmentally friendly with lower water and energy use. Berstorff's Klaus Koch, speaking at the 25th Microwave Power Symposium in Denver, Aug. 23, 1990, noted that a meal weighing 400 g requires 30–45 min processing time via heat conduction (retorting); the same meal takes only 3–5 min using a microwave system. Reduced thermal processing times correspond to improved product quality. The difference between pasteurization and sterilization is that pasteurization occurs at 165' to 190'F, while sterilization occurs at 250' to 265'F. Pasteurization reduces a product's bacterial counts, which leads to an increased refrigerated shelf life. Microwave pasteurization reportedly is more expensive than other pasteurization systems, but it offers advantages in processing times and operating costs.

Top Cuisine products from P&T Foods (the Netherlands) introduced six varieties including spaghetti in 1990 and had a line of prepared meals sterilized using microwave energy. These products were the first applications of commercial microwave sterilization process. The entrees were manufactured in round plastic trays.

Europe, however, enthusiastically embraced microwave treatment as a way to extend the shelf life of bread. Ever since the European Union came into existence, the regulations have prevented companies from adding preservatives to foods shipped across country borders. In order to ship bread, many bakeries use microwave pasteurization—particularly in Germany. The procedure simply involved running pallets of packaged bread loaves through a low-energy microwave tunnel. The treatment can extend the shelf life from around a week to 10 days up to two, three, or even four weeks.

After US FDA approval of MATS process, 915 Labs company (WA, USA) obtained the worldwide license to manufacture and market microwave-assisted sterilization and pasteurization technologies. According to 915 Labs, any food or beverage that will benefit from a lower processing temperature and a reduced processing time is ideal for MATS processing. Heat-sensitive foods such as eggs, dairy ingredients, seafood, and pastas have all been successfully processed with MATS. The reported research demonstrated that compared to conventional thermal retorting, the shortened exposure to heat during microwave processing retains more nutrients, such as Omega 3s, B vitamins, vitamin C, and folate.

The company manufactured a system that perform both sterilization (MATS) and pasteurization (MAPS) treatments. The pilot scale system was developed for product and process development. The smallest commercial production system processes 30 packages per minute. Larger capacity production systems with a throughput between 50 and 225 packages per minute are available by design. Actual capacities vary depending on the nature of the product, size of package, and the desired shelf life to be achieved. In order to realize the potential benefits of microwave processes and assist food companies to commercialize microwave technology and integrate it in the production process, 915 Labs offers product development, validation services, and packaging solutions.

5.8 Industrial radio frequency heating in processes and systems

As a rapid heating method, radio frequency (RF) heating offers a considerable speed advantage over conventional heating methods, particularly in solid foods in which heat transfer is mainly governed by heat conduction. Similar to microwave heating, RF dielectric heating is now widely used in industrial applications such as drying textile products, final drying of paper, final dehydration of bakery products at outlets of baking ovens, and melting honey (Wang et al., 2003). RF drying of food products has mainly been used for the postbaking drying of cookies, crackers, and pasta. The most successful applications often combine two or more drying techniques (heat pump, forced air with RF IR or microwaves) to improve drying system performance. It has been shown that the ability of RF energy to target the internal moisture content brings a significant performance improvement to a drying process. Cookies and crackers, fresh out of the oven, have a nonuniform moisture distribution that may give cracking during handling. RF technology heating assists to even out the moisture distribution after baking by targeting the remaining moisture pockets.

RF heating has been studied for the inactivation of bacteria in a variety of foods. Similar to microwave heating, a main challenge for commercial application of RF heating technology is in heating uniformity. RF technology has been explored in various food processing operations, such as pasteurization and sterilization and insect disinfestations in various commodities, such as fresh fruits and nuts, meat products, eggs and egg products, RTEs, dairy products, pasta, etc.

Heatwave Technology Inc (Mississauga, ON, Canada) validated RF heating for postpackage pasteurization of refrigerated RTE products. Pasteurization of cooked eggs, blanching of vegetables (broccoli and green beans), baking croissants, and drying of cookies were used as an example of the successful product treatment.

Another frequent application for RF pasteurization is in the treatment of dry ingredients. In a dry state (5%–15% moisture content), microbes are considered "dormant" and are challenging to kill. RF is effective with heating dry ingredients such as flour, cereal grains, protein supplements, spices, seeds, and pet foods, insuring their food safety.

Radio Frequency Company (MA, USA) has been marketing its Macrowave batch heating systems for the pasteurization and disinfestation of food products. A 30 kW, 40 MHz Macrowave Pasteurization System for granular food products such as green peanuts and various types of grain, which was designed and patented by RFC's founder, Mr. Joshua G.D. Manwaring on July 4, 1967. In addition, RF postbaking dryers have been utilized on cracker and snack food production lines to increase productivity and also eliminate checking problems. Recently, RF heating has grown into the preferred method to reduce water activity and also to provide a validated kill-step in the production of finished baked goods. In addition, RF system is implemented as a kill-step of Salmonella pathogen in pasteurization process of almonds and other dry ingredients. With the introduction of the FSMA, there is now a more drive to devise more applications of RF heating and validate it for the pasteurization of low moisture foods.

RF Food Systems (the Netherlands) is specialized in designing and developing applications for the food industry, especially in the sliced meats, snacks, and convenience food industry.

There are also applications in the wet pet food industry, for instance, in the production of chunks.

Two other companies manufacturing RF processing equipment for the food industry are Stalam Co., Italy and Strayfield Co., Berkshire, United Kingdom.

5.9 Modeling of microwave heating systems

The studies of microwave pasteurization and sterilization processes, in general, used systems not designed or optimized for these purposes due to the complexity of the microwave heating, where the heating pattern depends on such a large number of critical factors. Nonuniform and uncontrolled microwave coupling were a key problem during the majority of these research. Obviously, the design of a microwave processing system can dramatically influence the critical pasteurization or sterilization process parameter—the location and temperature in the coldest point in the tested product. This uncertainty in the published literature creates difficulties in the formation of general conclusions about microwave processing and its application for pasteurization or sterilization processes. As it was reported by IFT, US FDA, just two commercial systems worldwide currently perform microwave pasteurization and/or sterilization of foods.

Currently, computer modeling tools are more often used for a development of microwave heating–based preservation process for foods due to the complexity of interaction of EMF, food, and packaging. While the basic components of a microwave processing system generator, applicator, and control systems are standard, the interaction of food matrix and packaging with EMF and changes in fundamental food properties during heating make the development of microwave processes and validation very complex. This complexity may be dealt with the use of an integrated approach by a product developer, process design team including food engineer, the microwave equipment manufacturer, and an electromagnetic filed specialist.

Datta (1991) used mathematical modeling of microwave heating as a tool to study safety. He modeled microwave heating of solid slab and liquid such as tap water in two cylindrical containers of 11.2 cm (500 cc load) and 17.0 cm (2000 cc loads) in microwave oven cavity. The axial temperature profiles showed larger variations in temperature than in radial profiles. The cold points were at the bottom when microwaves entered radially. The experimental cold point was at the bottom. This was attributed to the oven design where microwave energy could not penetrate effectively through the bottom of the containers.

For the development of microbiologically safe processes, simulation-based design can save significant time and resources. This is because of the complexity of microwave systems, where the heating pattern depends on a large number of factors. A number of experimental techniques suggested as potentially capable of making the temperature field more uniform. This includes rotation of the load, mode stirrers, multiple feeds, sample movement, dielectric layers in the cavity, and design of the load and its container. Known modeling efforts are limited to simplistic Lambert's law-based calculation and 1D analytical-numerical and 2D finite element-based models.

As an example, modeling software, such as the QuickWave-3D has been used in the design of equipment for overcoming heating uniformity problems during pasteurization of liquid foods. Such software provides a comprehensive insight into the heating process by showing interior power absorption. The recent progress in computational electromagnetic engineering has resulted in a remarkable increase in the adequacy of modeling and accuracy of simulations and a notable growth of the use of computer simulations for accelerating the design and improving characteristics of wireless telecommunication equipment, semiconductors, computer systems, networking, and other products. The same trends have now become apparent in microwave power engineering.

Simulation-based design can be used to allow choosing proper combinations of food and process parameters in an efficient manner, reducing time and expenses for prototype building. Improved microwave system/product coupling characteristics and stability of operating regimes can be obtained as a result of simulation of EMF. Computation of the heating performance and selection of commercial microwave system parameters can be performed. Parameters of applicator, processing conditions (distribution of absorbed energy) have been selected by using one of the most powerful electromagnetic modeling tools available up to date, the finite difference time domain (FDTD) software *QuickWave-3D*.

The modeling can be a tool to resolve the questions such as:

1. Identify type of the microwave system that would be most appropriate for microwave heating of various products
2. Estimate the electric field in the chosen cavity for a particular type of field excitation
3. Determine distribution of the dissipated power and SAR within the processed product
4. Determine distribution of temperature within the cavity and product
5. Establish the minimum and maximum temperatures in the product and its temperature history

A general description of the analysis and synthesis computational tools of the modeling-based technique for solving the fundamental problem of intrinsic nonuniform microwave heating was described by Cordes and Yakovlev (2007). First piece of the software implements a FDTD solution of a coupled electromagnetic thermal problem with varying parameters and pulsing microwave energy. This tool allows for choosing the design variables for the optimization problem which is formulated as a minimization of time-to-uniformity and solved by the synthesis software producing a description of the optimal heating process with the resulting uniform 3D temperature field.

A novel temperature-based parameter for the uniformity of microwave-induced temperature patterns is introduced as the average squared temperature deviation corresponding to all distinct neighboring pairs of FDTD cells in the load. Computational experiments showed a substantial (up to orders of magnitude) reduction in time-to-uniformity and have exemplified by series of 3D temperature fields illustrating processes of forming more and more even temperature patterns. As an example of application, a computer modeling was used to design a slotted waveguide applicator for heating aqueous emulsion of polyvinyl chloride. Computer modeling was applied for a parametric study of geometric factors that can affect the system efficiency. The simulation allowed a far more effective optimization of the applicator design (Cresko and Yakovlev, 2003).

Computer simulations were used to solve the uneven microwave heating problem to design active packaging or packaging made could redistribute the electric field and thus generate uniform heat patterns within the product. The computer simulations were performed using *QuickWave*-3D 2.0 for a food sample in a rectangular paper box. The scenario was designed for the Daewoo 800 W 0.9 L domestic microwave oven. It was shown that the product was heated extremely nonuniformly. The top section was heated more uniformly, while the bottom was overheated. Computations conducted for the same cavity and the load in the absence of the package box showed practically no difference in the heating patterns. It was concluded that paper-based materials and other traditional packaging materials with low values of dielectric constant are not able to influence the electromagnetic field within the food (Yakovlev, 2001a, b).

5.10 Conclusions

The application of microwave domestic ovens and IMSs has more than 70 years history of development. There is microwave-based industrial operations such as cooking, drying, and tempering that demonstrated their commercial feasibility. Microwave-based pasteurization and sterilization obtained regulatory approvals and demonstrated high-quality products and may be the technology that will enable ecommerce providers to deliver high-quality food directly to consumers through their traditional distribution channels. The future of microwave processing of foods appears to be the strongest for special applications, and it will probably be of limited usefulness as a general method of producing process heat.

The advances in development of active microwave packaging and the growth success of application of computer simulation of microwave heating are potential ways to improve heating performance of microwave systems. However, there are not yet broadly established temperature measurement method and procedures in quality assurance, product and process development in the food industry. The further implementation of TTIs is promising for evaluating not just a process temperature but the process lethality.

Similar to microwave heating, RF dielectric heating is now widely used in industrial applications such as postbaking, drying products, tempering, pasteurization of low moisture products, and disinfestation of food products. The research results with RF heating over the past few decades have clearly shown that in the near future, RF will be a very attractive processing technology to provide safe and high quality of food products because of its ability to penetrate deeply with rapid uniform heating.

Another challenge is how to apply modeling theory and technology that will result in improvements of industrial microwave and RF processes. Presently, advanced modeling projects are mainly confined to universities studies. However, with rapid evolution in computers and software and the interdisciplinary character of microwave heating technology, there are major opportunities for the industry to take a lead in modeling applications.

References

AFFI (American Frozen Food Institute), 2008. Recommended Guidelines for the Labeling of Microwave Cooking Instructions. AFFI, Washington, DC.

Cordes, B., Yakovlev, V., 2007. Computational tools for synthesis of a microwave heating process resulting in the uniform temperature field. In: Proceedings of 11th AMPERE Conference on Microwave and High Frequency Heating (Oradea, Romania, September 2007), pp. 71–74.

References

Cresko, J., Yakovlev, V., 2003. A slotted waveguide applicator design for heating fluids. In: Proceedings of 9th AMPERE Conference on Microwave and High Frequency Heating, Loughborough, U.K., pp. 317–320.

Datta, A.K., 1991. Mathematical Modeling of Microwave Processing as a Tool to Study Safety. American Society of Agricultural Engineers. Meeting, p. 23.

Grocery Manufacturers Association. The Association of Food Beverage and Consumer Products Companies, 2008. Guidelines for Validation of Consumer Cooking Instructions for Not-Ready-To-Eat (NRTE) Products. Available online at: http://www.gmaonline.org/downloads/technical-guidance-and-tools/121894_1.pdf. Accessed in June 2022.

Guan, D., Gray, P., Kang, D.-H., Tang, J., Shafer, B., Ito, K., Younce, F., Yang, T.C.S., 2003. Microbiological validation of microwave-circulated water combination heating technology by inoculated pack studies. Journal of Food Science 68 (4), 1428–1432. https://doi.org/10.1111/j.1365-2621.2003.tb09661, 1267.

Kim, H.-J., Taub, I.A., 1993. Intrinsic chemical markers for aseptic processing of particulate foods. Food Technology 47, 91–99.

Kim, H.-J., Taub, I.A., Choi, Y.-M., Prakash, A., 1996. Principles and Applications of Chemical Markers of Sterility in High-Temperature-Short-Time Processing of Particulate Foods: ACS Symposium Series.

Schlegel, W., 1992. Commercial Pasteurization and sterilization of food products using microwave technology. Food Technology December 62–63.

Tang, J., Liu, F., March 17, 2015. Method for Recording Temperature Profiles in Food Packages during Microwave Heating Using a Metallic Data Logger. US8981270B2 inventors.

Tang, J., 2015. Unlocking potentials of microwaves for food safety and quality. Journal of Food Science 80 (8), E1776–E1793. https://doi.org/10.1111/1750-3841.12959.

Tang, J., Chan, T.V., Chow, T.C., Tse, V., 2007. Microwave and radio frequency in sterilization and pasteurization applications: heat transfer in food processing. In: Sunden, B. (Ed.), Heat Transfer in Food Processing: Recent Developments and Applications, pp. 101–142.

USDA-FSIS, 1999. Performance standards for the production of certain meat and poultry products. FSIS Directive 7111.1. Federal Register 64, 732–749.

US FDA Guidance Documents, 2007. Guidance for Industry: Preparation of Premarket Submissions for Food Contact Substances (Chemistry Recommendations). Accessed June, 2022. www.fda.govregulatory-information/search-fda-guidance-documents/guidance-industry-preparation-premarket-submissions-food-contact-substances-chemistry.

Wang, Y., Wig, T.D., Tang, J., Hallberg, L.M., 2003. Sterilization of foodstuffs using radio frequency heating. Institute of Food Technologists 68 (2), 539–544.

Wang, J., Tang, J., Liu, F., Bohnet, S., 2018. A new chemical marker-model food system for heating pattern determination of microwave-assisted pasteurization processes. Food and Bioprocess Technology 11 (7), 1274–1285. https://doi.org/10.1007/s11947-018-2097-2.

Yakovlev, V., 2001a. Commercial EM Codes Suitable for Modeling of Microwave Heating—A Comparative Review, Lecture Notes in Computational Sciences and Engineering. Springer Verlag (to be published).

Yakovlev, V., 2001b. Improving Quality of Microwave Heating by Packaging—Analytical Approach. ASAE meeting, Sacramento, CA. July 30-August 1, 2001.

CHAPTER 6

Economics, energy, safety, and sustainability of microwave and radio frequency heating technologies

6.1 Introduction

Commercially successful applications of microwave processing took advantage of characteristics unique to microwaves that are not available in the conventional processing of foods and other materials. Some of these key characteristics of microwaves' interactions with materials are:

- Penetrating radiation
- Controllable electric field distributions
- Rapid heating
- Selective heating or differential absorption of materials
- Self-limiting reactions.

These characteristics, either alone or in combination, represent opportunities and benefits not offered by conventional heating or processing methods. However, these obvious advantages also introduce problems and challenges that have to be addressed.

Table 6.1 summarizes some of the features, as well as the benefits and challenges, that are associated with each of the key microwave characteristics.

Commercialization of the microwave technology will be based not only on the areas of use and the properties of the materials produced but also on its economics. The cost of new processes must be considered at the same time as the technical aspects of microwave technology. The central question is "What are the key considerations in determining the costs associated with microwave processing of materials, and how do these costs compare with the costs of conventional processing?" An additional question that this review seeks to answer is "When is microwave processing most useful or appropriate?" Answering this question requires that the issue process economics would be addressed.

Table 6.1 Some key characteristics and features of microwave processing.

Characteristics of microwave heating	Features	Benefits over conventional heating	Disadvantages
Penetrating radiation, direct bulk heating	Materials heat internally Reversed thermal gradients from inside (ΔT) Lower surface temperatures Instantaneous power/temperature response Low thermal mass Applicator can be remote from power source	Potential to heat large sections uniformly Improved product quality and yields Materials and composite synthesis Automation, precise temperature control Rapid response to power level; pulsed power Heat in clean environment Differential coupling	M/W transparent materials difficult to heat Hot spots, cracking Large ΔT in low thermal conductivity materials and nonuniform heating Controlling internal temperature Arcing, plasmas Require new equipment designs special reaction vessels
Field distributions can be controlled	High energy concentration Optimize power level versus time	Precise heating of selected regions (brazing, welding, plasma generation, fiber drawing) Process automation, flexibility, energy saving	Equipment more costly and complex Requires specialized equipment
Dielectric losses accelerate rapidly above T_{cnt}	Very rapid heating	Shorter processing time (2–1000x factor) Heat materials >2000°C Capable to heat M/W transparent materials > T_{cnt}	Hot spots, arcing Nonuniform temperature distribution Control of thermal runaway
Differential coupling of materials	Selective heating of internal or surface phases, additives, or constituents	Heating of microwave transparent, materials via additives, fugitive phases, etc. Hybrid heating (active containers) Materials synthesis Selective zone	Reactions with unwanted impurities Contamination with insulation or other phases

Table 6.1 Some key characteristics and features of microwave processing.—*cont'd*

Characteristics of microwave heating	Features	Benefits over conventional heating	Disadvantages
Self-limiting	Selective heating ceases (self-regulating) after certain processes have been completed	heating (joining, brazing, and sealing) Controlled chemical reactions, oxidation, reduction; use of microwave transparent containers Below critical temperature, drying and curing are self-regulating Completion of certain phase changes is self-regulating	Undesired decoupling during heating in certain products Difficult to maintain temperature

The important consideration is that the economic feasibility is a function of local variations in energy costs, environmental laws, and labor costs balanced with the properties of finished materials or parts, improvements in yield or productivity, and the markets' availability for the products. This chapter therefore seeks only to discuss the various cost aspects, provide guidelines for what must be considered, provide costs or savings as appropriate or available, and give examples as available. A definitive accounting of the costs cannot be given because industrial microwave technology is still not fully commercialized, the range of application is broad, and the costs-benefit ratio will always be product specific.

It should be noted that microwave processing is unlikely to be economically competitive with processing using natural gas in the foreseeable future because of the difference in costs between natural gas (approximately $6.50 per MBTU) and electric power (approximately $17.50 per MBTU). The values cited are typical energy costs for industrial applications. Actual energy costs vary regionally.

The intrinsic performance characteristics of microwave heating have been discussed earlier in this chapter. Krieger (1989) has suggested to consider the following characteristics of products and processes that may potentially make them attractive for microwave processing.

- The size or thickness of the material should be large.
- Improvements in properties or quality reachable from microwave processing are significant.
- Plant space is limited.
- Electricity is economical.

- Minimizing handling is advantageous.
- Heat from the combustion of coal, oil, or natural gas is not practical (i.e., electricity is the only power source).
- Maintaining a very clean, controlled processing environment is important.

The cost issues examined in this chapter include: cost of capital equipment, including comparison with conventional equipment; operating costs (energy, replacement, maintenance); energy required per part (energy efficiency) and cost of energy; deletions and addition of steps from conventional processing schemes; savings in time and space and changes in yield over conventional processing. In addition to economics of commercial microwave heating, the safety aspects for humans and sustainability aspects of food processing and packaging for consumers, food manufactures, and environment will be also discussed in comparison with traditional heating methods because of their key importance in future developments of science and technology.

6.2 Cost of capital equipment

The equipment used for industrial microwave processing is generally custom designed and optimized based on specific application needs. As described in previous chapters, the basic components of a microwave heating system include: the power supply, magnetron power source, followed by cavities and applicators, feed systems, seals, and controls and sensors.

The cost of microwave equipment depends on size, power rating, frequency, applicator design, gas control system, peripherals, manufacturer, and the size of the market for that particular equipment. Because of these dependencies, capital costs vary widely depending on the applications. Typical cost ranges are presented in Table 6.2 (Sheppard, 1988). Due to the differences in the configuration and processing approach between microwave and conventional systems, it is very difficult to perform a general comparison of capital costs in a meaningful way. However,

Table 6.2 Capital cost of industrial microwave equipment.

Component	Typical cost
Complete system	$1000–5000/kW
Generator	<50% of system cost
Applicator	>50% of system cost
Power transmission	($1000–3000), <5% of system cost
Instrumentation	($1000–3000), <5% of system cost
External materials handling	($1000–3000), <5% of system cost
Installation and start up	5–15% of system cost

Source Sheppard, L. M., 1988. Manufacturing ceramics with microwaves: the potential for economical production. Ceramic Bulletin 67 (10), 1556–1561.

microwave processing equipment is almost always more expensive than conventional systems.

Information regarding microwave generator costs is given in Table 6.3.

6.2.1 Operating costs

Operating costs include the cost of energy, both absolute and the real cost per part based on the coupling efficiency and the size and number of parts, and the cost of maintenance, repair, and replacement. Table 6.4 gives some estimates of these costs.

6.2.2 Energy efficiency and cost

The overall energy efficiency of microwave heating systems is defined by Lakshmi et al. (2007) as the product of the microwave generation efficiency and the microwave absorption efficiency. Microwave generation efficiency largely depends on the quality of the microwave power source (i.e., magnetron), while absorption efficiency depends on the level of impedance matching between the source and the load (processed food)—which depends on the load's dielectric properties, mass, geometry, and cavity. Microwave output power is based on real-time measurements rather than prior manufacturers' ratings and settings. The microwave power outputs of domestic 2450 MHz microwave ovens have traditionally been calibrated based on the IEC 60705 standards. The calorimetry-based calibration method (IEC 60705, 1999) involves heating 1000 g $\pm$5 g of water (m_W) inside a cylindrical borosilicate glass container (m_C, external diameter 190 mm, height 90 mm) from T1 = 10°C $\pm$1°C to T2 of 20°C $\pm$ 2°C.

$$P_{water} = \frac{(4.187\ m_w\ (T_2-T_1) + 0.55\ m_c\ (T_2-T_0))}{t} \quad (6.1)$$

t is heating time, s

Microwave power sources have a maximum generation efficiency of about 70% (at 2450 MHz) and 88% (at 915 MHz), whereas overall is in the range of 25%–55% (Atuonwu and Tassou, 2019). Atuonwu and Tassou (2017) analyzed the energy implications of nonuniform heating by evaluating the maximum percentage energy savings associated with improving the heating uniformity of microwave processes. Depending on specific processing conditions, energy savings in the order of 20% were shown to be possible for pasteurization and sterilization processes, while about 85% savings are achievable for thawing processes where the nonuniformity is severe. Control of the source and load variables is needed to minimize reflections or maximize absorption efficiency. Magnetrons have poor flexibility and feedback control. Solid-state microwave power sources have proved capable of receiving feedback on the dynamic states of the load and controlling the source power delivery

Table 6.3 Microwave tube characteristics.

	Power KW	Frequency GHz	Efficiency %	Cost $K	Cost $/Watt	Special requirements[a]
Magnetron						
"Cooker"	1	2.45	60–70	0.05	0.05	
Industrial	5 to 15	2.45	60–70	3.50	0.35	
Industrial	50	0.915	60–70	5.00	0.10	
Power grid (transmitting)						
"Low"	To 10	To 3	80s		0.1–0.4	Ex
"Medium"	To 100	To 1	80s		0.10	Ex
"High"	To 2000	To 0.15	80s		0.10	Ex
Klystron						
Example	500	3	60s	350.00	0.70	S
Example	250	6	60s	200.00	0.80	S
Example	250	10	60s	375.00	1.50	S
Klystrode						
Example	30	1	70s	18.00	0.60	S, ex
Example	60	1	70s	36.00	0.60	S, ex
Example	500	1	70s	Dev.	~1	S, ex
Gyrotron						
Example	200	28	30s	400.00	2.00	Sc
Example	200	60	30s	400.00	2.00	Sc
Example	500	110	30s	500.00	1.00	Sc
Example	1000	110	30s	800.00	0.80	Sc
Example	1000	140	30s	Dev	~1	Sc

ex, external circuitry, cavities, etc; S, solenoid for beam focus; Sc, superconducting solenoid.
[a] Special requirements.

Table 6.4 Operating costs of microwave equipment.

Component	Typical cost
Magnetron replacement	1–12 ¢/kW·h
Electric energy	5–12 ¢/kW·h
Plug-to-product efficiency	
915 MHz	70–75%
2450 MHz	50–65%

to match with the needs of the load. Atuonwu and Tassou (2019) specified the advantages of solid state over magnetrons including:

- Accessibility to feedback and process optimization algorithms
- Possibility of multiple power feeds with independent control
- Frequency control, accuracy, and stability
- Phase control
- Power amplitude control
- Repeatability and reproducibility of product quality
- Reliability and lifetime
- 15 years continuous operation against about 500 h in magnetron-powered household ovens, and one year of continuous use in commercial and industrial environments
- Size and operational voltages—about 50V or less compare to the 4–20 kV of magnetrons, at least, 10 times lighter for the same wattage
- Improved compatibility with other electronic circuitry and the Internet-of-Things in the future

In terms of costs, solid-state generators are still more expensive. For the cheapest technology laterally diffused metal-oxide semiconductor, $0.25/W is now achievable in some cases as against >$1.00/W in the past. Magnetrons cost $0.01/W to $0.1/W (Edgar and Osepchuk, 2001). It is, however, noteworthy that magnetron-based commercial/industrial systems have a limited lifetime of one year.

In analyzing energy costs for microwave processing, it is important to consider how energy is used in such operations. The theory of microwave generation and materials interactions are discussed in Chapters 1 and 2. A simplistic view is presented here only for illustration. Input power is that generated by the magnetron. However, the power absorbed by the component depends on the coupling characteristics of the component, the number and arrangement of components in the cavity, and the cavity design. A certain amount of power is reflected. The absorbed energy is the difference between the input and the reflected power. It is clear, therefore, that improvements in coupling efficiency of the load material and the arrangement of insulation and components have significant effects on the energy efficiency. In typical commercial applications, microwave processes have overall efficiencies of 50—70% (Metaxas and Meredith, 1983).

Energy savings arising from the use of microwave energy should be considered on the basis of the energy cost for the entire process. Hybrid systems that combine conventional and microwave heating are beneficial for many applications. Drying, for example, is best approached by first removing the bulk of the water by conventional means and then removing the remainder rapidly using microwave heating.

6.3 Savings from processing changes

The greatest potential for microwave processing is in increased productivity and a consequent decrease in labor, rejection, and space costs. If improvements in properties over conventionally processed materials are realized, the premium in price obtainable for such improvements should not be forgotten. There have been a number of reports of savings in time and improvements in productivity obtained by microwave processing of foods that are summarized in Table 6.5.

In pasteurization and sterilization processes, the focus of interest, according to Ohlsson (2001) has been on the processing of packaged, precooked foods. Experimental work with finished cooking and pasteurization of packaged prebrowned meat patties by microwave processing showed that a shelf life of five to six weeks at chilled temperature storage was obtainable after 2 min of microwave heating to reach 80°C. It was claimed that in Europe, over a 100 pasteurization plants of at least four different equipment manufacturers have been installed over five years according to this review. They claim microwave heat time about 5 min after preheat to ~40°C to the desired final temperature of 80—85°C. A typical pasteurization plant for 2 tons/h consists of up to eight hybrid waveguide/cavity chambers with a total of some 200 low-power magnetron modules.

The economic benefits of microwave processing are difficult to define in a general way. The decision to use microwave processing for any application has to be based on an analysis of the specific process. Important factors include the location of the processing facility; the product requirements; possible property improvements; alternative sources of energy; availability of capital; regulatory requirement and the balance between energy costs, labor costs, capital costs, and the value added to the product (Edgar, 1986).

For example, the Multitherm concept for sterilization has been based on many years of fundamental and applied research on microwave sterilization at the Alfa Laval Company and SIK (Sweden). This has comprised the determination of dielectric data in the high temperature region, computer simulation work, and experimental pasteurization and sterilization work. It has included quality studies on sensory and nutritional aspects as well as inoculated pack studies to assure microbiological safety of the products. The quality of the microwave-sterilized vegetables was found significantly higher than that of the conventionally processed sample and equivalent to the frozen products. A high degree of uniformity of heating is needed to achieve a superior quality in microwave sterilization process (Ohlsson, 2001).

Table 6.5 Examples of improvements for microwave-processed foods.

Material	Process	Time savings	Productivity improvement	Performing organization	Source
Bacon cooking	Cooking	20%–40%	Increased yield, improved properties, less energy use, improved shelf life Incoming product 2,000Lb/h Finished product 800 l b/h 40,000 slices per hour	Microdry 25 systems	Shiffman (1992)
Meat tempering	Raising temperature from 0 to 25 F	Reduced cycle to 5 min;	Space savings, high throughput 15,000 l b/hr	Raython 200+ systems	Shiffman (1992)
Pasta drying	Moisture removal	Shortens drying time from 8 to 0.5 h	Improved sanitation and quality	Microdry 10–20 systems	Shiffman (1992)
Bread baking	Heating	Shortens time from 22 min to 6–8 min	Efficiency of energy utilization	Continental baking Co	Shiffman (1992)
Pasteurization of RTE meals	Heating to destroy pathogenic and spoilage m/o	5 min from 40 to 80°C	Two tons/hour Better quality product Shelf life extension up to 5–6 weeks	100+ MW units in Europe	Ohlsson (2001)

6.4 Safety of microwave heating

Microwave radiation can heat human body tissue the same way it heats food. Exposure to high levels of microwaves can cause a burn. Two areas of the body, the eyes and the testes, are particularly vulnerable to radio frequency heating because there is relatively little blood flow in them to carry away excess heat.

Through the Center for Devices and Radiological Health (CDRH), the FDA sets and enforces standards of performance for electronic products to assure that radiation emissions do not pose a hazard to public health. The US Federal Standard (21 CFR 1030.10) limits the number of microwaves that can leak from an oven throughout its lifetime to five mW of microwave radiation per square centimeter at approximately two inches from the oven surface. This limit is far below the level known to harm people. Microwave energy also decreases dramatically as you move away from the source of radiation. A measurement made 20 inches from an oven would be approximately 1/100th of the value measured at two inches from the oven.

The standard also requires all ovens to have two independent interlock systems that stop the production of microwaves the moment the latch is released or the door is opened. In addition, a monitoring system stops oven operation in case one or both of the interlock systems fail.

IEEE Standard C95.1-1991 recommended power density limits for human exposure to radio frequency and microwave electromagnetic fields. In the radio frequency—microwave frequency range of 100 MHZ to 300 GHZ, exposure on the limits are set on the power density (W/cm^2) as frequency. The recommended safe power density limit is as low as 0.2 mW/cm^2 at the lower end of this frequency range because fields penetrate the body more deeply at lower frequencies. At frequencies above 15 GHZ, the power density limit rises to 10 mW/cm^2, since most of the power absorption at such frequencies occurs near the skin surface.

Using patented applicator designs such as chokes, microwave ovens, and other commercial equipment manufacturers is able to reduce electromagnetic leakage from system entry and exit points to practically nondetectable levels in microwave heating chambers. Chokes are used around doors, windows, and seams to prevent microwave leakage. This poses no threat of electromagnetic energy to the health and safety of equipment operators if all precautions are followed. As a further precaution, control systems are supplied with safety interlocks and leakage detectors that shut down power instantaneously in the event of equipment malfunction.

Modern pacemakers are shielded from stray electromagnetic forces and have a backup mode that takes over if a really strong electromagnetic field disrupts the main circuit's programming.

Another myth is arching and the noticeable sparks that are visible when a metal object is placed inside an operating microwave device. In fact, microwaves can be applied to products in metal vessels as long as the mass of product and the intensity of microwaves are matched. Arching takes place when too much energy is applied to a given mass of product in the vessel.

6.5 Sustainability of microwave and radio frequency heating and equipment

With growing consumers awareness, in particular, in developed countries, food and its production are an essential measure of healthy life, it is clear that sustainability is a key driver for the future evolution of agri-food primary sector and processing. The key principles of sustainable food manufacturing that can reduce their environmental footprint include:

- Reduced food waste and loss: Food and Agriculture Organization of the United Nations (FAO UN) defined food waste as "the decrease in quantity or quality of food." Thus, food waste is part of food loss and refers to discarding or alternative (nonfood) use of food that is safe and nutritious for human consumption along the entire food supply chain. According to the FAO, nearly a third of food produced for human consumption gets wasted annually.
- Improved energy and water efficiency: Processing and manufacturing are responsible for about 23% of the food industry's overall energy costs in the United States. Water is used in food processing, both as an ingredient, and in various industrial processes for cleaning, sanitizing, heating, and cooling media.
- Use of sustainable additives, ingredients to improve quality value, and degradable or recyclable packaging that is often a major source of waste and pollution.
- Achieving circular economy.

According to these principles, sustainable food processing is about innovative ways to meet present needs without comprising future viability in changing economic and environmental conditions. Furthermore, to be sustainable, a technology must not be harmful to either the environment, animals, or people. The impact of modern radio frequency and microwave technologies on environmental, economic, and social aspects of human activity as the basis of sustainable development should be discussed.

Application of innovative electromagnetic heat processing technologies for the replacement of conventional heat processes could provide a potential to reduce current levels of energy and water consumption, production costs, and improve the sustainability of the food sector without the infrastructural changes of production chains meaning that sustainable food processing is the only approach for the industry in order to succeed.

In the case of food processing, microwave heating is applied for microbial reduction to achieve improved safety and longer shelf life of prepared products with higher quality, nutritional and sensorial attributes. Additionally, microwave technology is generally described as a sustainable approach due to the low-energy consumption and short-time processing. Multiple differences of microwave heating in comparison to conventional thermal treatments indicate that in food production, microwaves can be applied as a technology combining product and packaging development and processing. As a processing method, microwaves are effective against pathogens and spoilage organisms in RTE products. As a process for new product development, microwavable product design targets the less changes in physical properties including proteins and visual appearance. Two main paths for microwaves application are foreseen: (1) fast reheating of RTE products in domestic applications; (2) numerous commercial batch and continuous operations where faster heating results in better products and higher productivity.

Both paths of microwaves application can lead to more sustainable food processing, packaging, and product development. First of all, microwave heating contributes to social sustainability by providing high-quality, nutritious, safe, and healthy food for consumers with an acceptable shelf life in developed countries. Correct choice of microwave processing and packaging conditions can result in improvements in the taste, nutrients content, and texture of food products compared to traditional processing methods. The potential for microwaves and radio frequency technologies to create longer shelf life for refrigerated or chilled foods also opens the opportunity for popular food items to be moved from frozen to chilled processing and distribution—and thus to reduce energy consumption.

Second, microwave heating could be simultaneously applied for a range of raw and semifinished products and allow food processors to reduce or eliminate ingredients added solely for preservative effects, including chemicals that inhibit bacterial growth. Reduced ingredient usage means eliminating the environmental impacts associated with sourcing those ingredients. This is a major reason that microwaves and radio frequency technologies can be considered as a clean label in the industry and for consumers.

Next, microwave heating generally causes no marked nutritional loss in foods compared with conventional thermal processes and increases foods' shelf life at refrigerated storage, thus reducing food loss and waste. Economically, food loss and food waste reduction allow easy access to food by decreasing food prices. From food waste, different types of compounds such as phenolics and antioxidants can be extracted. As an effective heating technique, microwaves can be used for improving the extraction of natural extracts from food waste.

Additionally, electromagnetic heating technologies have less impact on the environment as they are based on utilizing strictly electrical energy, no water, or steam and have lower CO_2 emissions than conventional thermal processes. Both methods require less energy than chilling and freezing. Research indicated that microwave heating processes consumed less energy per kg of food than conventional thermal processes.

Protecting food and agricultural products from insects is a serious problem in many countries. The use of radio frequency energy allows to replace chemical treatment of the agricultural products and grains by methyl bromide or other fumigants and eliminate the environmental risks.

Therefore, microwave and radio frequency treatments are considered consumer and environmentally friendly compared to conventional heat processing.

6.5 Sustainability of radio frequency heating and equipment

6.5.1 Microwave packaging sustainability

The combination of sustainable microwave heating and sustainable microwave packaging enables sustainability in packaging and processing science. Numerous sustainable materials have been researched and employed to realize sustainable packaging. Bio-based polymers: polylactic acid (PLA), polyhydroxy butyrate (PHB), and poly(3-hydroxybutyrate-co-3-hydroxyvalerate) (PHBV) show promising characteristics for many microwave applications. PLA has been widely investigated for food packaging applications because of its excellent transparency and relatively good water resistance. The drawbacks of PLA material include low thermal and shape stability, brittleness, and low melt strength. Disposable cups, lids, and straws are examples of commercial PLA products used for cold foods and beverages applications, and the appearance of PLA cups is similar to that of regular polystyrene (PS) plastic cups. PHB polymer has been widely investigated for food packaging applications due to its good moisture resistance and barrier properties. Similar to PLA, PHB is brittle and its thermal stability is low. PHBV plastic is studied to use for food packaging application because of its good moisture resistance and barrier properties but also due to easier processing ability compared with that of PHB. However, thermal stability of PHBV is relatively low. For successful utilization of such ecofriendly and biodegradable materials for microwave food packaging, their thermal and dimensional stability should be improved. Modification of these bio-based polymers by using nanotechnologies and emerging technologies has been suggested to provide superior and smart microwave packaging based on ecofriendly materials for convenience foods (Thanakkasaranee et al., 2022).

6.5.2 Domestic microwave ovens

Fast technological developments and falling prices are driving the purchase of microwave domestic ovens worldwide. Consumers also tend to buy new appliances before the existing ones reach the end of their useful life as electronic goods. Discarded electronic goods are one of the fastest growing waste streams worldwide. A further issue associated with their growing consumption is the increase of household electricity consumption over the last decade. The growing number of domestic microwave ovens as kitchen appliance has a significant impact on the environmental, economic, and social aspects of human life.

The comprehensive environmental evaluation of microwave domestic ovens across the whole life cycle assessment and implications at the EU level of the implementation of the Standby regulations were assessed by Gallego-Schmid* et al. (2017). The system boundaries of the study comprised the following stages:

- Production of materials such as galvanized steel, aluminum, brass, copper, ferrite, tin, lead, gold, nickel, silver, zinc, and palladium; plastics: acrylonitrile butadiene styrene, polyethylene terephthalate, polyvinyl chloride, polypropylene, glass fiber reinforced nylon, PS, and polyoxymethylene; tempered glass; ceramics; and cardboard (for packaging);

- Manufacturing of microwaves: metal cold impact extrusion and plastic molding; production of electronic components; steel powder painting and curing, product assembly and packaging;
- Use of microwaves: consumption of electricity;
- End-of-life waste management: disposal of postconsumer wastes;
- Distribution: transport of microwave materials and packaging to the manufacturing facility, microwave ovens to retailer and end-of-life waste to the waste management plan.

It was reported, using the microwave would consume 537.6 kWh of electricity over its lifetime of eight years (1200 cycles/yr at 0.056 kWh/cycle); the reference microwave oven (1200 W) will use 8.9 GJ of primary energy and contribute around 416 kg CO_2 eq. over its service life. The analysis suggested that electricity utilization in the usage stage is the main contributor to the environmental impacts. Additionally, the resources to manufacture the microwave ovens are the major contributors to the exhaustion of abiotic elements (67%) and play an essential role for human and marine toxicities as well as formation of photochemical oxidants (13%–27%). Manufacturing of microwaves also contributed to 10 out of the 12 environmental impact categories considered in the review, especially photochemical oxidants (34%), ozone layer depletion (33%), and acidification (31%). End-of-life removal had an overall positive effect on the environmental impacts due to the recycling, in particular, for depletion of elements and human toxicity.

The authors recommended that in order to improve the environmental footprint of microwave ovens further, it is necessary to develop specific regulations for these devices, considering other life cycle stages apart from the use. Possible future trends, such as shorter microwave lifetime, electricity decarbonization, and reduced availability of some materials, will make it even more critical to understand better the effect on the impacts of other life cycle stages, such as production of materials, microwave manufacture, and waste treatment. Improvement in these parts of the supply chain would result in significant environmental benefits, particularly as the number of microwave ovens is expected to increase in the future. They concluded that the development of a specific eco-design regulations for microwaves, focused on all the life cycle stages, should be an objective for the near future, and additional research should be centered on analyzing the potential for eco-design of microwaves.

6.5.3 Microwave-assisted technologies to achieve circular economy

The circular economy is the concept linked to the reduction, reuse, and recycling of food losses and wastes along the food supply chain. According to a survey conducted by FAO (2011), the estimate is that about one-third of the food produced in the world for human consumption is lost or wasted, which represents approximately 1.3 bln tons per year. Achieving reduced food waste and losses involves identifying the dynamics of bioresidues generation and developing mechanisms to avoid this generation, as also managing unavoidable losses through strategies such as reuse of by-products. There are a number of bioresidues that can be a valuable source for the recovery of high value-added compounds, mainly with antioxidant capacity. The replacement of preservatives and other additives from artificial origin, usually related to adverse effects on human health, faces some challenges such as availability and cost. Opportunities to obtain these compounds can be found in the food industry itself because a great variety of food waste has been identified as a source of high value-added compounds. Large amounts of seeds, fibrous strands, peel, bagasse, among other parts of fruits and vegetables are lost or wasted during industrial processing, despite being rich sources of bioactive compounds. From a circular economy perspective, bioactive compounds, such as phenolic compounds, can be largely recovered mostly from seeds and peels, and successfully incorporated back into foods. These bioactive compounds (e.g., antioxidants, polyphenols, tocopherols, carotenoids, and vitamins) naturally present in food are important for human nutrition. When these compounds are incorporated into the human diet, via food products or medicinal bases, they can provide benefits to health, developing positive effects on body functionality.

6.5.4 Microwave-assisted extraction

Conventional methods generally require long extraction times to obtain greater performance and involve large amounts of solvent, which are sometimes toxic. Alternative and eco-friendly extraction techniques, including microwave heating, have showed advantages in extracting antioxidant and preservatives compounds in comparison with conventional methods. Alternative methods are more sustainable, and more ecological techniques with high extraction yield and compounds preservation.

Microwave-assisted extraction (MAE) is a method that uses microwave heating. A solvent is used for compound extraction from a sample placed in the microwave zone where the biomolecules and solvent align with the alternating microwave fields. The larger the dielectric constant of the solvent, the higher the heating and dipoles rotation. Through microwaves internal heat generation mechanism, pressure is generated inside the plant cell and causes its consequent rupture, exposing the cell and then facilitating solvent penetration. The operation of MAE involves three successive steps: releasing the solutes present in the active sites of the plant matrix under increased temperature and pressure; diffusion of solvent through the plant matrix; and finally, solute release, cellular content, from the matrix to the solvent.

The extractable yield of pectin was 13.85% from mango peel using MAE at 700 W oven for 3 min, and the product had greater porosity compared to the conventional method (Wongkaew et al., 2020). A continuous-flow microwave system was designed to extract the pectin from potato pulp by Arrutia et al. (2020) using 2,450 MHz, and 400 or 800 W for 5—60 min under constant stirring at 600 revolutions per minute. The yield of pectin after extraction was 40%—45% that was more than twice that from using water extraction. Reduction of extraction time and energy consumption was also reported by Jesus et al. (2019), by using the MAE method in comparison to the conventional heat-assisted one. In addition to these advantages, through the alternative method, the authors achieved higher yields and concentrations of ellagic acid.

Grape pomace is the main winemaking by-product of the wine chain generated after the pressing of grapes in the production of white wine and/or after the fermentation stage in red winemaking. Grape pomace is rich in phenolic compounds, which possess several physiological effects, directly linked to their medical, technological, antioxidant, and antimicrobial properties. The phenolic compounds present in grapes is classified into phenolic acids (hydroxybenzoic and hydroxycinnamic acids), flavonoids (catechins, flavonols, and anthocyanins), and proanthocyanidins. Polymeric polyphenols (proanthocyanidins) present in grape peels and seeds are often described as tannins (or condensed tannins), while monomeric flavonoids, present in red grape peels, are known as anthocyanins. MAE is one of the most promising techniques to extract and conserve phenolic compounds and reuse grape pomace because of the demonstrated benefits in reduction of the extraction time, number of unit operations, energy consumption, environmental impacts, costs, quantity of solvent and waste production, and achieving high quality extracts. Alvarez et al. (2017) used the MAE as a pretreatment operation and found 57% of improvement in the extraction efficiency of polyphenols from grape pomace. Parameters such as choice of solvent, solvent/solid ratio, applied power and temperature, and extraction times must be considered using the MAE method. The individual or combined effects on the yield of the extracts and their composition are considered fundamental factors to maximize the extraction process.

In summary, advantages and limitations of MAE include a lower cost equipment, which requires reduced extraction time and quantity of solvent (ethanol or water), with improved extraction yield. MAE also allows processing without using solvent. On the other hand, the heat generated can affect heat-labile compounds and it loses efficiency in scaling up (Leichtweis et al., 2021).

6.5.5 Microwave-assisted pyrolysis of food waste

Pyrolysis is a process of decomposition where biomass undergoes thermal degradation in the absence of oxygen or other oxidants. This is a practical method for producing value-added products such as biochar, biooil, and biogas from biomass such as food waste. In recent years, more research explored applications of various microwave heating pyrolysis processes such as microwave vacuum pyrolysis, fixed-bed microwave pyrolysis, or microwave flash pyrolysis for the conversion purposes. The interest in this process caused by obvious advantages of microwave pyrolysis include shorter reaction time, higher heating rate, heating selectivity, energy saving, and fewer pollutants emission. Also, microwave pyrolysis has some disadvantages such as need in absorbents for organic materials with poor polarity, higher capital costs, and challenges with control of heating rates. In contrast, traditional heating pyrolysis is more advanced and frequently used in pyrolysis studies and processes.

Overall, it is considered that microwave pyrolysis is a promising technology to replace the traditional landfills and incineration of food waste disposal and convert it into value-added products which have future applications and benefits. Numerous research has been focused on synthesis of biochar from food waste using microwave ovens and optimizing processing conditions (temperature in the range of 200–500°C, time, output power, N_2 or CO_2 gas flow) to obtain necessary porous structure of the product. Such studies demonstrated that using microwave energy sunflower husks were converted into activated carbon, sugarcane bagasse and orange peel were converted into biochar at the significantly faster rates than conventional heating in biochar production. Moreover, many studies found that microwave pyrolysis can produce larger biochar surface area than traditional heat process, which should be attributed to the reverse temperature gradients in microwave process. In food waste pyrolysis process, the microwave produced biochar surface area ranges from 419 to 1511 m^2/g, while traditional heating is from 404 to 1794 m^2/g. This can be attributed to the emission of gaseous products inside the particles that could leave rich pore structures. Also, in the case of the microwave pyrolysis of sunflower husk, it was observed that the biochar pores were aligned more uniformly compared to the char obtained using traditional heating.

Biooil has been considered as an alternative energy source of traditional fuels and chemical raw materials because it is cleaner and renewable. Biooil was produced using fast pyrolysis and flash pyrolysis methods. Nowadays, microwave pyrolysis for producing biooil received more interest. Waste cooking oil, wasted plastic, different types of food waste, and deoiled fish waste were used as raw material sources for pyrolysis of biooils. Maximum biooil yields were obtained at 500–700°C in both microwave and traditional pyrolysis process. As temperature or microwave power increased, the biooil yields were significantly influenced or reduced, which should be attributed to the secondary carking (aromatization, polymerization, and carbonization) of the condensation. Moreover, the optimal temperature for biooil yield in microwave pyrolysis was lower than that under traditional heating, reported as 350–581°C (MP) and 500–900°C (THP) that can be considered as microwave process advantage.

Through pyrolysis process, the conversion of food waste into biogas can be also achieved. Biogas, as a renewable energy source, can originate from waste and many other forms of biomass, and it can be used for combustion to generate power. Food waste, coffee hull, pomegranate peel, tea bags, kitchen waste, and other types of agri-food production were investigated as sources for biogas production. In general, it has been concluded that microwave pyrolysis is a rapid and energy-saving method. By applying suitable catalysts, this microwave energy advantage could be more significant. Due to the high temperature (400–600°C) and heating rate, microwave pyrolysis is more suitable for biooil production. For biogas, high temperature (>700°C) and long pyrolysis time are favorable. Moreover, using metal oxides catalysts to improve the final temperature and heating rate in microwave pyrolysis could also be considered in biogas production. According to Li et al. (2022), nonuniformity and various sources of food waste make challenges in food waste pyrolysis.

Compared to agricultural wastes, food waste microwave pyrolysis requires more support and research investment due to its advantages. Its industrialization is a challenge due to complexity in reactor designs, process control, and security. Furthermore, most microwave pyrolysis research has been mainly performed in lab-scale units, against a growing need for large-scale disposal operations.

6.5.6 Next development steps

Even though there are some limitations for the current applications due to the higher cost and uniformity issues, microwave heating sustainability benefits make this technology considerable and competitive for numerous technological applications on the market now and in the future. It is still an emerging technology, which discovers these new but advanced applications. Further research is needed in terms of cost optimization and scalability to fulfill the needs of the industry, environment, and the consumer. Also, the engineering design of commercial microwave equipment and microwave domestic appliances should target the advances toward improved efficiency of energy consumption for food processing and developments of food waste reduction.

6.6 Conclusions

The commercialization of microwave processing is hindered by the high capital costs of microwave systems and the inherent inefficiency of electric power. In most successful industrial uses of microwaves, factors other than energy account for savings realized from microwave processing; improvements in productivity and material properties; and savings in time, space, and capital equipment are probably the best bases for selecting microwaves over conventional processes. In many applications, hybrid systems provide more savings than either microwave or conventional systems on their own.

Factors affecting energy efficiency include power source, load, and source-load matching factors. This highlights the need for highly flexible and controllable power sources capable of receiving real-time feedback on load properties and effecting rapid control actions to minimize reflections and maximize absorption, heating non-uniformities and other factors that lead to energy losses. From results of the economic analysis, solid-state sources are promising replacements to magnetrons not only from an energy and overall technical perspective but also in terms of economics. Although magnetron generators are fairly cheap, solid-state microwave generators have several advantages over vacuum tubes: they deliver precise control of the frequency, power, and phase of the radio frequency signal, high reliability, low input voltage requirements, and compactness.

Advances in the microwave processing of solid, liquid, and semiliquid substances and materials have proved that this low energy-consuming technology can be considered a sustainable manufacturing technology. Continuing improvements in microwave and radio frequency equipment heating efficiency, cycle time and productivity, and application of biogradable packaging materials will further reduce energy requirements and contribute to sustainability of food production. Microwave and radio frequency energy significantly reduces the time of thermal treatment. This allows increasing productivity in the industry and also accelerates some chemical reactions.

The selectivity of radio frequency and microwave heating, when only the sample absorbs energy, allows saving input power. The use of only electrical energy reduces the use of fossil fuel in the industry and chemicals in agriculture. The compactness of microwave equipment and lower overall operating costs are an attractive alternative to traditional heat treatment methods and systems. From results of the economic analysis, solid-state sources are promising replacements to magnetrons not only from an energy and overall technical perspective but also in terms of economics. All these aspects of technologies development can promote sustainable economic growth and improve living standards. Over the next decade, microwave technology is expected to become even more popular, especially in developing countries.

More opportunities should be researched for the role of microwave technology in pursuing and shaping a circular economy. The successful applications of microwave energy in food waste reduction and loss in the operation such as microwave extraction or microwave pyrolysis are confirmed by numerous research. More efforts focused on technological progress are also needed to integrate these processes into the solution systems.

Separation of resource consumption and environmental impacts of microwave units from the economic growth is essential, and improvements in microwave equipment are a key product category that can contribute to it. The efforts to reduce the environmental impacts of higher electricity consumption should be combined with the development of ecologically friendly designs of microwave ovens that stipulate the optimization of resource use. Possible future trends, such as shorter lifetimes and limited availability of some resources, make the development of such product along with harmonized worldwide regulations more critical.

References

Alvarez, A., Poejo, J., Matias, A.A., Duarte, C.M.M., Cocero, M.J., Mato, R.B., 2017. Microwave pretreatment to improve extraction. efficiency and polyphenol extract richness from grape pomace. Effect on antioxidant bioactivity. Food and Bioproducts Processing 106, 162–170. https://doi.org/10.1016/j.fbp.2017.09.007.

Arrutia, F., Adam, M., Calvo-Carrascal, M.A., Mao, Y., Binner, E., 2020. Development of a continuous-flow system for microwave-assisted extraction of pectin-derived oligosaccharides from food waste. Chemical Engineering Journal 395, 125056, 10.1016/j.cej.2020.125056.

Atuonwu, J.C., Tassou, S.A., 2019. Energy issues in microwave food processing: a review of developments and the enabling potentials of solid-state power delivery. Critical Reviews in Food Science and Nutrition 59 (9), 1392–1407. https://doi.org/10.1080/10408398.2017.1408564.

Atuonwu, J.C., Tassou, S.A., 2017. Quality Assurance in Microwave Food Processing: A Review of Developments and the Enabling Potentials of Solid-State Power Delivery.

Edgar, R., 1986. The Economics of Microwave Processing in the Food Industry, June, pp. 106–112.

Edgar, R., Osepchuk, J.M., 2001. Consumer, commercial and industrial microwave ovens. In: A, K., Datta, R.C., Anantheswaran (Eds.), Handbook of Microwave Technology and Food Application. Marcel Dekker, New York, pp. 215–277.

FAO, 2011. Global Food Losses and Food Waste-Extent, Causes and Prevention. United Nations, Rome, Italy, pp. 1–37.

Gallego-Schmid*, A., Mendoza, J.M.F., Adisa Azapagic, A., 2017. Environmental assessment of microwaves and the effect of European energy efficiency and waste management legislation. Science of The Total Environment 618, 487–499. https://doi.org/10.1016/j.scitotenv.2017.11.064.

IEC 60705, 1999. Household Microwave Ovens—Methods for Measuring Performance. International Electrotechnical Commission. IEC Publication, Geneva.

Jesus, M.S., Genisheva, Z., Roman, A., Pereira, R.N., Teixeira, J.A., Domingues, L., 2019. Bioactive compounds recovery optimization from vine pruning residues using conventional heating and microwave-assisted extraction methods. Industrial Crops and Products 132, 99–110, 2019.

Krieger, B., 1989. Marketing Combination Heating Systems. Presented at Industrial Heating Workshop. International Microwave Power Institute. November 1989.

Lakshmi, S., Chakkaravarthi, A., Subramanian, R., Singh, V., 2007. Energy consumption in microwave cooking of rice and its comparison with other domestic appliances. Journal of Food Engineering 78 (2), 715–722.

Leichtweis, M.G., Oliveira, M.B.P.P., Ferreira, I.C.F.R., Pereira, C., Barros, L., 2021. Sustainable recovery of preservative and bioactive compounds from food industry bioresidues. Antioxidants 10, 1827. https://doi.org/10.3390/antiox10111827.

Li, H., Xu, J., Nyambura, S.M., Wang, J., Li, C., Zhu, X., Feng, X., Wang, Y., 2022. Food waste pyrolysis by traditional heating and microwave heating: a review. Fuel 324 (2022), 124574. https://doi.org/10.1016/j.fuel.2022.124574.

Metaxas, A.C., Meredith, R.J., 1983. Industrial Microwave Heating. Institute of Electrical Engineers. Peter Peregrinus, Ltd, London.

Ohlsson, T., 2001. Microwave technology and foods. Advances in Food and Nutrition Research 43, 66–140.

Sheppard, L.M., 1988. Manufacturing ceramics with microwaves: the potential for economical production. Ceramic Bulletin 67 (10), 1556–1561.

Shiffman, R., 1992. Microwave processing in the US industry. Food Technology 50–53.

Thanakkasaranee, S., Sadeghi, K., Seo, J., 2022. Packaging materials and technologies for microwave applications: a review. Critical Reviews in Food Science and Nutrition. https://doi.org/10.1080/10408398.2022.2033685.

Wongkaew, M., Sommano, S.R., Tangpao, T., Rachtanapun, P., Jantanasakulwong, K., 2020. Mango peel pectin by microwave-assisted extraction and its use as fat replacement in dried Chinese sausage. Foods 9, 450. https://doi.org/10.3390/foods9040450.

CHAPTER 7

Conclusions, knowledge gaps, and future prospects

Microwave and radio frequency (RF) heating is a promising advanced food processing technology that has been developing for more than 70 years. The advantages of microwave and RF heating are in their ability of rapid volumetric heating that is desirable for delivering premium foods with higher quality, nutritional and sensory attributes. Also, microwave and RF technology offer:

- increasing processing productivity because of reduced treatment time
- higher energy efficiency of food operations
- energy-saving opportunities
- more environmentally friendly than conventional heating modes.

In this regard, microwave heating has found increasing interest for developing and application in a wide variety of domestic, commercial, and industrial food processing operations. This extensive end use, any energy, and other savings in microwave-based food processing operations can have a large positive impact on food processing sustainability.

The current use and future of microwave processing of foods appears are on the rise but appeared to be strongest for special purposes, and it will probably be of limited usefulness as a general method of producing process heat. The successful commercial applications include tempering of frozen vegetables, meat, fish, seafood, and poultry products, and precooking of bacon for foodservice and cooking of sausages. Other processing operations and processing aids include postprocessing drying, cooking, preheating, extraction, disinfestation, pasteurization, and sterilization of liquid foods and beverages, prepared and cooked solid ready-to-eat prepackaged meals, and low and intermediate moisture products and ingredients, seeds, and nuts.

In order to facilitate development and acceptance of microwave technology, future research efforts have to be directed toward improving microwave process performance in terms of its uniformity, streamline process validation, packaging, and microwave equipment improvement through broader use of solid-state power sources instead of magnetrons.

Further research is necessary on microwave effects on quality and nutrients, with more emphasis on moisture effects, protein, carbohydrate, and vitamin retention and chemical reactions. A collaborative effort by researchers using microwave heating mode has to be made to standardize experimental procedures, measure and report all heating characteristics of the systems, packaging of the products, and to minimize research variability of data. The development of microwave cooking techniques be pursued to optimize food quality. Microwave penetration, shapes, and geometry of packaged products must be taken into account to avoid overheating.

Lack of reliable temperature measurement and control during microwave heating of foods has discouraged efforts to assess temperature distribution and history, locate the least heated spot in the product, and measure microbial inactivation kinetic and nutrients destruction that consequently resulted in complexity and challenges of microwave pasteurization and sterilization process validation. The industrial time-temperature integrators using chemical, enzymatic, or biological indicators have to be developed and used instead of temperature probes or in combination with the probes because they present a practical cost-effective way to evaluate process lethality value in the load volume.

Because food packaging materials and containers can affect the microwave heating process, more focus should be made on the development of active packaging for solid and semisolid multicomponent products. There are promising advances and innovations in the development and application of active packaging and understanding of material behavior that became a part of microwave heating process. Implementation of aseptic packaging is a promising approach to apply for liquid products and beverages with extended shelf life.

Computer simulations have to become an essential tool to solve the uneven microwave multiphysics heating problem to design a process considering interactions with food product, active packaging or packaging, and cavity that can redistribute the electric field and thus generate more uniform temperature and heating patterns within the product.

Index

'*Note:* Page numbers followed by "f" indicate figures and "t" indicate tables.'

A

Absorbing materials, 44
Acrylamide formation, 105–106
Active food packaging, 115–116
Aluminum trays, 116
American Frozen Food Institute (AFFI), 123–124
Ascorbic acid destruction, food quality, 90

B

Bacon precooking and crisping, 113
Beef frankfurters, 71–72
Biogas, 153–154
Biooil, 153
Blanching, 15
Boneless meatball cooking, 113
Browning agents, 14

C

Capacitive dielectric heating, 10
Capital equipment cost, 140t
    configuration and processing approach, 140–141
    energy efficiency and cost, 141–144
    microwave heating safety, 146
    operating costs, 141
    processing changes and savings, 144
Carbohydrates, food quality, 96
Cavity magnetron, 30
Circular economy, 150–151
Commercial applications, 157
Commercialization, 137, 154
Computer simulations, 158
Conductivity, 39
Contact sensor, 34
Continuous fluidized bed microwave paddy drying system, 16–17
Conventional cooking, 113
Conveyor-belt-tunnel applicators, 32
Corn-soy–milk (CSM) food blend, 84
Coupling coefficient, 36
Crystallized polyester containers, 117

D

Debye resonance, 49
Domestic microwave ovens, 149–150
Drying process, 113

E

Economic feasibility, 139
Electrical conductivity, 40
Electric resistant heating, 4
Electromagnetic heating technologies
    advantages and limitations, 13t
    heat exchangers, 3–4
    infrared (IR) heating, 6, 6f
    magnetic induction (MI) heating, 5
    microwave heating, 3–4
    ohmic (OH)/electric resistant heating, 4, 4f
Energy efficiency, 154
Enzymes destruction, food quality, 104t
    α-amylase destruction, 102t
    Bacillus subtilis-amylase (BAA), 99, 101f
    blanching, 99
    browning reactions, 99
    D- and z-values, 101
    horseradish POD destruction, 99
    orange juice, 102, 103t
    pectin methylesterase (PME), 102
    thermal stable enzyme, 99
    wheat germ lipase and soybean lipoxygenase inactivation, 101
Explosion puffing process, 18

F

Fiber-optical sensors, 120
Fluidized bed drying, 16–17
Foodborne and spoilage microorganism destruction
    beef frankfurters
        athermal microwave effects, 72
        computer-controlled microwave heating system, 71–72
        microwave heating system, 71
        pasteurization/sterilization, 71
        polyvinylidene chloride (PVDC) film, 72
        typical thermal process, 71
    bottled pickled asparagus, 71
    chemical marker method, 56–57
    cold point, 56
    energy absorption, 56
    energy sources and chemical preservatives, 73
    food properties
        acid foods, 64–65

Index

Foodborne and spoilage microorganism destruction (*Continued*)
  chemical composition, 65–66
  composition, 64
  intrinsic factors, 64
  low-moisture foods, 65
  roduct pH and thermal process, 64
  Salmonella species, 66
  thermal resistance, 64
  water activity reduction, 65
inactivated microorganisms, 56
in-package sterilization and pasteurization, 70
kinetics data, 57
macaroni and cheese product, 70–71
microbial inactivation kinetics
  conventional thermal processing, 58
  decimal reduction time, 60
  D-value, 59–60, 60t
  F-value, 59
  microwave preservation process, 57–58
  nonlinear heat transfer model, 60
  nth order chemical reaction, 58
  thermal resistance, 59
microwave continuous flow systems
  juices, 66–68
  milk and protein beverages, 68–69
microwave energy
  inactivation effects, 62t–63t
  log kill data, 61
  microwave-resistant foodborne pathogens, 61
  Z-values, 61
nonuniform heating, 56
pulsed microwave (PW) radiation, 73–74
salsa, 70
sweet potatoes, 69–70
temperature measurement and control, 56–57
Food packaging materials and containers, 158
Food quality
  acrylamide formation, 105–106
  carbohydrates, 96
  chemical reactions
    browning and good texture, 103–105
    jet impingement, 105
    Maillard reaction, 103
    toxic compounds, 105
  continuous-flow microwave heating, 102
  enzymes destruction, 99–103, 104t
  food processors, 82
  lipids, 94
  low glycemic index foods, 98
  microwave oven, 82
  microwave pasteurization, 82–83
  microwave preservation, 82
  minerals, 98
  nonuniformity, 81
  overall quality
    color and taste, 85–86
    corn-soy–milk (CSM), 84
    flavor, 86–87
    foil pouch-packed foods, 83t
    food composition components, 84
    microwave high-temperature short-time (MHTST) sterilization, 83
    moisture content, 85
    packaging and process temperature, 83, 84t
  polyphenols, 91–93
  polysaccharides, 96–97
  proteins, 95–96
  ready-to-eat (RTE) meal product quality, 82
  starch, 97–98
  sterilization, 82
  thiobarbituric acid (TBA) values, 106–107
  time-temperature history, 81
  vitamins and nutrients destruction
    ascorbic acid, 90
    folacin losses, 88
    inactivation kinetics parameter, 89t
    reported effects, 92t–93t
    riboflavin destruction, 89
    thiamine, 87
    two reheating methods, 90–91
    vitamin B1, 89
    vitamin C destruction, 88–90, 91t
    vitamins B_{12}, 90
    vitamins E and B1, 89–90
    water-soluble vitamin, 88
Freeze drying, 16
Frying process, 17–18

G

Glycemic index (GI), 98
Grape pomace, 152
Grocery Manufacturer's Association (GMA), 123–124

H

Heating characteristics
  coupling coefficient, 36
  dielectric properties and specific heat capacity, 36
  domestic and commercial microwave heating systems
    industrial microwave systems, 34–35
    magnetron, 30

microwave ovens, 32–33
smart microwave oven, 33–34
solid-state generators, 30–31
waveguides and applicators, 31–32
flow regime, 37
food electrical conductivity, 40
foods dielectric properties
   capacitivity, 38
   chemical composition, 42
   conductivity, 39
   dielectric loss factor, 38
   microwave heating, 39
   moisture content, 41–42
   nonelectrolytes in water, 43
   organic solids, 43
   permittivity, 38
   pH and ionic strength, 43
   proteins, 43
   radio frequency range, 49–50
   relative permittivity, 38–39
   tangent of dielectric loss angle, 38
   temperature, 41
heating efficacy, 35
incident microwave power, 35
microwave electromagnetic wave propagation
   absorbing materials, 44
   dielectric properties, mixtures, 46–49
   foods transmission properties, 44–45
   reflecting materials, 44
   wave impedance and power reflection, 45–46
microwave heating rates, 36
microwave power absorption, 40
product density, 40
product geometry, 37–38
spatial temperature distribution, 36
specific heat, 40
time-temperature heating curves, 36
High-density polypropylene (HDPP), 115

I

Incident microwave power, 35
Industrial microwave heating systems, 34–35, 114
   multimagnetron tunnel ovens, 127
   ready-to-eat meals and in-pouch sterilization
      Berstorff Corp., 129
      Classica Microwave Technologies Inc, 128
      continuous sterilization process, 128–129
      equilibration, 127–128
      fresh filled pasta, 128
      holding and cooling time, 127–128
      915 Labs company, 130

      microwave preheater, 127–128
      microwave sterilizer, 127–128
      preservatives, 129
Industrial radio frequency heating systems
   dielectric heating, 130
   dry ingredients treatment, 131
   drying system, 130
   Heatwave Technology Inc, 131
   Macrowave Pasteurization System, 131
Industrial time-temperature integrators, 158
Infrared drying (IRD), 21–22
Infrared heating food processing operations, 7f
   conveyorized IR broiling, 8
   electromagnetic radiation spectrum, 6f
   hotdogs, 8
   maximum radiation, 6–7
   microbial inactivation, 7
   operating efficiency, 8
   ready-to-eat (RTE) fully cooked meat contamination, 8
Intrinsic performance characteristics, 139–140
Ionic drift, 11–12

J

Jet impingement, 14, 105

L

Lycopene content, 86

M

Macaroni and cheese product, 70–71
Magnetic induction (MI) heating, 5, 5f
Magnetron, 30
Maillard reaction, 103
Microwave and radio frequency bands
   dielectric heating mechanism, 9
   electric displacements, 11–12
   food processing operations
      baking, 14–15
      blanching, 15
      drying, 15–17
      freezing, 21
      frying, 17–18
      infrared baking, 21
      infrared drying, 21–22
      infrared heating, 21
      infrared roasting, 22
      infrared tempering, 22
      microwave-assisted extraction (MAE), 18
      ohmic heating, 22–23
      pasteurization and sterilization, 19–21
      puffing, 18–19

Microwave and radio frequency bands (*Continued*)
  tempering and defrosting, 19
  food processors, 10
  frequency allocation, 10t
  heat generation mechanisms, 11–12
  magnetron applicator, 8
  microwave energy generation, 9
  microwave products, 10
  temperature distribution, 12f
Microwave-assisted extraction (MAE), 151–152
  advantages and limitations, 152
  grape pomace, 152
  operation, 151
  pectin yield, 151–152
  solvent, 151
Microwave-assisted thermal pasteurization (MAPS), 19–21
Microwave-assisted thermal sterilization (MATS) process, 19–21
Microwave heating process, 3–4
Microwave high-temperature short-time (MHTST) sterilization, 83
Microwave ovens
  inhomogeneous heating pattern, 32–33
  invention, 32
  mode stirrer, 32–33
  nonionizing radiation, 33
  turntable, 32–33
Microwave processing characteristics, 138t–139t
Microwaves' interactions, 137
Microwave vacuum frying (MVF), 17
Molded pulp containers, 117
Multimagnetron tunnel ovens, 127

N

Near-field applicators, 31
Noncontact sensor, 34
Nonself-venting materials, 117–118
Not-ready-to-eat (NRTE) foods, 123
Nutrient retention, 55–56

O

Ohmic heating, 4, 4f
  benefits, 4–5
  devices, 4
  direct OH conduction losses, 4
  liquid products, 4–5
  microwave-assisted, 22–23
Operating costs, 141
Osmotic dehydration, 17

P

Packaging process
  active packaging, 115–116
  material properties, 114
  material reaction, 114
  microwavable foods and cooking instructions
    classification, 123
    food processors guidelines, 123–124
    microwave oven, 122–123
    microwave oven validation tests, 124–125
    not-ready-to-eat (NRTE) foods, 123
  microwavable packages
    crystallized polyester containers, 117
    flexible packages and pouches, 116
    molded pulp containers, 117
    nonself-venting and self-venting materials, 117–118
    paper board containers, 117
    rigid plastic tray, 117
    safe coated aluminum tray, 116–117
  modeling tools
    computer modeling tools, 132
    computer simulations, 133–134
    FDTD solution, 133
    mathematical modeling, 132
    preservation process, 132
    QuickWave-3D, 132
  passive packaging
    glass and glass jars, 115
    physical strength, 115
    plastic materials, 115
    requirements, 115
    volatile compounds, 115
  regulatory status and commercialization, 126
  temperature and process lethality measurements
    process lethality indicators, 121–122
    temperature probes, 120–121
    thermal imaging, 121
  validation
    Clostridium botulinum, 119
    documentary evidence, 118
    heating parameters, 119
    heating uniformity, 118–119
    Listeria monocytogenes, 119
    shelf life studies, 119
    spatial temperature distribution, 119
Paper board containers, 117
Passive food packaging process
  glass and glass jars, 115
  physical strength, 115

plastic materials, 115
   requirements, 115
   volatile compounds, 115
Pasteurization process, 55
Pectin methylesterase (PME) destruction, 102
Peroxidase (POD) and texture testing, 71
Polyphenols, food quality, 91–93
Polysaccharides, food quality, 96–97
Proteins, food quality
   biological process, 95
   denaturation, 95
   hydration, 96
   plant proteins, 95
   soybeans, 95
Puffing process, 18–19
Pulsed microwave (PW) radiation, 73–74
Pyrolysis
   biooil, 153
   disadvantages, 152–153
   microwave-assisted, 152–154
   traditional heating, 152–153

R

Radio frequency heating, 10, 11f
   food industry applications, 23–24
   food processing operations, 23–25
   insect infestation, 25
   tempering, 24–25
Rancidity, cottonseed oil, 94
Resistant starch (RS), 98
Riboflavin destruction, 89
Rigid plastic tray, 117
Roasting process, 22

S

Salsa, 70
Self-venting materials, 117–118
Single-mode applicators, 32
Slowly digestible starch (SDS), 98
Smart microwave oven, 33–34
Solid-state generators, 30–31
Starch, food quality, 97–98
Sustainability
   circular economy, 150–151
   domestic microwave ovens, 149–150
   electromagnetic heat processing technologies, 147
   improved energy and water efficiency, 147
   microwave-assisted extraction (MAE), 151–152
   microwaves application paths, 147–148
   nutritional loss, 148
   packaging, 149
   pyrolysis, 152–154
   reduced food waste and loss, 147
Sweet potatoes puree (SPP), 69

T

Temperature measurement and control, 158
Tempering process, 19, 113
Thawing process, 19
Thiobarbituric acid (TBA) values, 106–107
Time/temperature-indicators (TTIs), 121–122

V

Vacuum drying, 16
Vacuum frying (VF) technique, 17
Vitamin B_1 degradation, 89
Vitamin C destruction, 88–89

W

Waveguides and applicators, 31–32
Wave impedance and power reflection, 45–46

Printed and bound by CPI Group (UK) Ltd, Croydon, CR0 4YY
08/06/2025
01896869-0016